Stephen Lilley

Transhumanism and Society

The Social Debate over Human Enhancement

 Springer

Stephen Lilley
Department of Sociology
Sacred Heart University
Fairfield, CT
USA

ISSN 2211-4548   ISSN 2211-4556 (electronic)
ISBN 978-94-007-4980-1   ISBN 978-94-007-4981-8 (eBook)
DOI 10.1007/978-94-007-4981-8
Springer Dordrecht Heidelberg New York London

Library of Congress Control Number: 2012942026

Printed on acid-free paper

Springer is part of Springer Science+Business Media (www.springer.com)

In gratitude to my parents and Mary
For Sean, Ian, Brenna & Arianne

Preface

From the stories of deities and superheroes we imagine what it is like to be strong, impervious, and immortal. However, we know that the gulf between them and us is wide. The gods take serious their birthright and guard their gifts, and we will never duplicate the strange twist of fate that created the comic book legends. As we ponder the differences we cannot help but be reminded of our human limitations. We respond to mortality with acceptance, despair, or defiance. Perhaps, what first caught my attention about transhumanity was its pure expression of defiance. Transhumanity is about emancipation and transcendence through science, engineering, and technology. It is a bold plan to alter the human condition. It is a possible future that we should take seriously.

This book is not a tribute or critique of transhumanity, instead it is an introduction to the social debate over enhancing body and mind. I present this controversy as it unfolds in the contest between transhumanists and conservationists. The former are recognized to be the vanguard of scientific utopianism. Their movement has been gaining ground over the past two decades. The conservationists make strange bedfellows, from the religious right to the secular left, but they push back with an argument to conserve human nature and to ban enhancement technologies.

I hope to reach readers interested in the social and ethical implications of technological advancements. I identify the key contested points and present the debate in an orderly, constructive fashion. This book engages the reader in the discussion over humanism, the tension between science and religion, and the interpretation of socio-technological revolutions. I trust the reader to form his or her own judgments.

I am a sociologist and I draw insights from social movement literature as well as science and technology studies (STS). I treat the transhumanity debate as a call to arms by which the contestants strive to mobilize support to shape policies and institutions. Accordingly, I describe rhetorical strategies in terms of social movement recruitment and political campaigning. Drawing on STS, I offer an analysis of science and technology claims and counterclaims.

Throughout this project, colleagues have been most supportive. I would like to thank Fran Grodzinsky, Tom Curran, Jeff Cain, and Shirley Jackson for their reading of the manuscript. Jerry Reid and Joshua Klein offered valuable responses to my presentation on transhumanity for the Faculty Scholarship Seminar Series at Sacred Heart University in 2009. Parts of Chap. 4 were presented that year at the 4th International Conference on Technology, Knowledge and Society in Boston, MA and at the Annual Meeting of the Society for Social Studies of Science (4S) in Washington D.C. Part of Chap. 5 was presented in 2007 at the 2nd International Conference on Interdisciplinary Social Sciences in Granada, Spain. I am grateful to members of the audiences for their helpful comments and suggestions. Special thanks go out to those who attended my first talk on this subject back in 2005 at the 28th Annual Northeast Popular Culture Association (NEPCA) Conference in Fairfield, CT. Their enthusiastic reception encouraged me to broaden the scope of my study. Finally, in the past few years I have invited students to discuss and debate transhumanity. I have been so impressed with their thoughtful remarks. I am confident that this generation will be able to successfully negotiate the challenges ahead.

Contents

Chapter 1
Introduction to the Transhumanity Debate

In 2006 the Edge Foundation invited over one hundred distinguished scholars to respond to this prompt suggested by Steven Pinker: "The history of science is replete with discoveries that were considered socially, morally, or emotionally dangerous in their time; the Copernican and Darwinian revolutions are the most obvious. What is your dangerous idea?" (Edge 2006) The responses at www.edge.org are fascinating but let us consider for the moment that there are dangerous ideas and expand the set to include any ideas, not just from science, that break out and threaten established ways of thought and social norms. In their time, monotheism, predestination, manifest destiny, Nietzsche's overman, racial purity, and social darwinism did just that. String together a few dangerous ideas and you have a dangerous ideology, for example, the last few concepts were key components of Nazism. The political philosophy behind democracy brings together the dangerous ideas of equality, liberty, and universal rights.

The ideologies mentioned above, especially discredited Nazism, no longer have the element of surprise, their novelty mitigated by exposure and time. So, what dangerous ideologies today have the power to shock? I believe that transhumanism fits the bill. The "singularity," "natural-born cyborgs," and "proactionary principle" (all described later) are provocative ideas and the clarion call to use technologies to alter the human species makes transhumanism most daring. Opponents go so far as to claim that this ideology poses a grave threat to humanity.

The first article of the Transhumanist Declaration at the Humanity+ website [the doing-business-as name for the World Transhumanist Association (WTA)] holds that "Humanity stands to be profoundly affected by science and technology in the future. We envision the possibility of broadening human potential by overcoming aging, cognitive shortcomings, involuntary suffering, and our confinement to planet Earth." (Humanity+ 2010) In a previous statement the WTA declared: "We support the development of access to new technologies that enable everyone to enjoy **better minds, better bodies** and **better lives**. In other words, we want people to be **better than well**" (World Transhumanist Association 2008, original emphasis).

S. Lilley, *Transhumanism and Society*, SpringerBriefs in Philosophy,
DOI: 10.1007/978-94-007-4981-8_1, © The Author(s) 2013

The transhumanists offer a hopeful vision at a time when things could be better, should be better. We face interminable war, food and water shortages, global warming, economic instability, senseless violence, and many of us have little faith that current social institutions can handle these problems. Contrast the daily feed of bad news with upbeat headlines about promising breakthroughs from science and medicine, for example, "Newly Created Microbe Produces Cellulose and Sugars for Biofuels," "Gene Therapy Restores Some Vision In Nearly Blind Patients," and "Microchip That Can Detect Type and Severity of Cancer Created." I readily admit that after hearing about another bombing, mass shooting, casualties from the drug war, etc., etc., the technology stories give me a lift. Rather than showing humanity for being stupid, destructive, and juvenile, these stories provide examples of intelligence, creativity, and maturity. Transhumanists assert that this is what we do best as a species. Why shouldn't we embrace our strengths and use technologies to uplift the human race?

It is a seductive message, nevertheless critics counter that after the twentieth century with its gas chambers, chemical and nuclear warfare, and environmental pollution, no responsible person should accept without question the belief that we will find deliverance through science and technology. Corporations and the military, mainly interested in profit and weapons, sponsor much research and development. Technological disasters occur and fixes backfire. Why place humanity in greater jeopardy? I believe that we must be as willing to face such doubts as we would embrace hope, and a good place to start is the transhumanity debate.

Presenting the Transhumanity Debate

I will be using the label, "transhumanists," as a catchall for a variety of notable figures, many of whom accept this descriptor and all of whom advocate human enhancement. "Conservationists" is the term that I have chosen for their opponents because of their stand to conserve human nature and institutions. One of the more fascinating aspects of this social debate is the diversity within this opposition camp. Social conservatives, theologians, secular humanists, and environmentalists have found common cause in their preference for conservation. For instance Bill McKibben, a noted environmentalist, admires humans for their natural endowments and does not wish to see us compromised by artificial enhancements. Others want to conserve a time-honored way of life, a traditional understanding of *Imago Dei* (image of God in human), or the liberal state.

Around the turn of the millennium, influential scholars representing faith and secular traditions began to single out transhumanists for criticism. Ray Kurzweil, multi-millionaire inventor and entrepreneur and tireless promoter of human transcendence, received the most attention. He has written best sellers on the topic and in 2009 released the movie, *The Singularity Is Near*. He is the Chancellor of the Singularity Institute. Its mission is to prepare leaders to understand and manage advancing technologies for the greater good. Kurzweil intends to live long enough to reach the point when aging and disease will be conquered and to this end he partnered with Terry Grossman, M.D., to develop an anti-aging regimen of a specialized diet, supplementals, and chemical i.v. treatments. They have marketed a line of longevity products and have written a health book, *Transcend: Nine Steps to Living Well Forever* (2009).

Daniel Lyons (2009) begins his *Newsweek* article on Ray Kurzweil, *I, Robot: One Man's Quest to Become a Computer*, by citing Kurzweil's willingness to become a cyborg and ends the article this way:

> He has no doubt. None. He is utterly, completely, 100 percent sure that he is going to live forever. He will be reunited with his beloved father, and they will become immortal and spend eternity together. Nothing can talk him out of it. And that, at the end of the day, may be the scariest, or saddest, thing of all (73).

What I did not find in Lyon's article was an appreciation for the controversy that with great energy, determination, and skill Kurzweil has helped create. This is the problem when too much attention is placed on personalities. As I see it, this was a missed opportunity. Science and technology controversies abound but often they entail disputes over specific methods, facts, or effects. Not so the transhumanity debate– it is expansive both with regard to the wide range of contested issues and the overt discussion of values and beliefs. Even though the central question to this debate is whether to accept or reject technological enhancement, the contestants address this question according to their respective metanarrative of what we are and what we should become and of what constitutes the greater good. They argue about human nature, ethics, and the proper role of government. The culture wars loom and rifts are evident between secular and religious and between proponents and opponents of global capitalism. It is an interesting story!

I chose to study the transhumanity debate for two reasons, one professional and the other personal: (1) As a sociologist, I wanted to investigate a sweeping science and technology controversy—none better than this one ostensibly about the future of humanity. For this investigation I primarily relied on concepts and methods from Science & Technology Studies (STS) and the social movement literature. (2) As a parent, I wanted to deepen my own thinking about the enhancement question to better deal with a family matter. I suppose that sooner or later many parents will face the enhancement question head on. My wife and I wrestled with a decision whether or not to have our twin sons undergo human growth hormone (HGH) treatment to increase their height. Using C. Wright Mills' [(1959) 1969] terminology, this was a *private trouble* for my family. Following Mills' advice I gained a better understanding by taking a look beyond my family to study the *public issue*. I found the transhumanity debate to offer the most comprehensive discussion on human enhancement and I appreciated the diversity of perspectives and the quality of the insights. However, I also found it to be sprawling and, at times, overly polemical. I set out to identify the main substantive points and to present the debate in an orderly, constructive fashion.

Christopher Tindale asserts that

> It is often in the contrast between positions, clearly and fairly presented that we come to
> understand what is at stake in each. In a world of compelling social debates so resistant to
> resolution, where intelligent well-meaning people stand on both sides of an issue, then this
> kind of understanding is not just the best we might expect, it is good in itself (2004:185).

I agree with Tindale and I hope that my presentation will be as helpful for the
reader as the exercise of preparing it has been for me.

In this first chapter I provide an overview of the transhumanist and conservationist
positions. In the second chapter I present the transhumanists' case for transcendence
and the conservationists' warnings of transgression. The opposing stands regarding
whether and to what extent we should transform the mind and body are described in
Chap. 3. There is more to the transhumanity debate than the juxtaposition of
principles and ideas, utopian and dystopian visions. The transhumanists are not
content to simply discuss the merit of enhancement, rather they are working to build a
world favorably aligned. Conservationists strive to prevent this. The transhumanity
controversy is as much a political act as it is an intellectual exercise. I treat the
transhumanity debate as a call-to-arms by which the contestants strive to mobilize
support to bring about change in policies and institutions. Accordingly, I identify
rhetorical strategies just as analysts do when studying social movement recruitment
and political campaigns. This is most apparent in Chap. 4 on the rhetoric of risk and
Chap. 5 on the claims of inevitability. In the final chapter, I offer some thoughts on
how the debate will end.

Transtechnologies and Society

Besides a preoccupation with personalities, another way to stymie an investigation
of the transhumanity debate is to focus exclusively on the technologies. We do well
to adhere to the STS admonition not to treat technology as an independent force.
Of course, without biotechnology, neurotechnology, information technology,
nanotechnology, and the associated sciences, transhumanism would not exist. The
Latin prefix, "trans" denotes crossing over, and the transhumanists must be able to
point to something that may serve as a likely bridge for humanity's passage.
Science fiction writers offer tantalizing sketches of futuristic technologies and they
pose very interesting "what-if" scenarios. However, you can't promote a dangerous
ideology, let alone a social movement, on imaginary technologies such as warp
drives.

Fortunately for those who yearn for radical change, biotechnology and the other
advanced technologies have become so well established in the public eye that they
can be enlisted to lend credence to visions of transcendence and transformation.
It is important to note that even opponents of transhumanism perceive these
technologies to be very powerful. The level of institutional support, for one thing,
must be recognized. These technologies are backed by well-financed research

programs and directed by the best scientists and engineers in the world. Corporations and nation states compete fiercely over them. The military is interested in certain applications. The media, as I mentioned previously, bestow lavish attention. Certainly there is no history of neglect or decline to suggest that these technologies will fizzle away.

Transhumanists and their critics also understand these technologies to have the capacity to intervene or to interfere (depending on one's perspective) in life at the most fundamental level. Most of us are familiar with this idea as it pertains to biotechnology. DNA and related genetic structures are regarded as the code of life, "cracked" by scientists and now open to manipulation through engineering techniques. Stem cells, basic in their pluripotency, have been coaxed from embryos, placentas, and skin and are being primed to promote regeneration.

Neuroscience is expected to get at the root of consciousness, emotions, and addictions. In regard to nanotechnology, Robert Frodeman explains that it's power

> turns on smallness in general, and the characteristics and possibilities that result from diminished size. At the nano-scale the physical behavior of matter changes; new material possibilities become attainable, for instance, extraordinary strength and lightness. Our relationship to matter itself is changed as items can be built from the ground up, atom by atom and molecule by molecule. Lost is the brute givenness that had previously characterized our relationship with nature. Certainly, we have long been able to manipulate materials, fashioning carbon or gold into a variety of items, but we had not attempted to refashion carbon itself. Now the prospect exists of changing carbon into gold, or vice versa (2006:384–385).

At the most basic level, so the thinking goes, transmutation is possible. Understandably, Frodeman and others see these technologies as the new alchemy.

With that said, we risk misunderstanding the debate if we assume that it is all about the technologies. The contestants are really arguing over sociotechnical ensembles. (Bijker and Pinch 1984) Let me explain. A crude causal model holds that technology determines society, e.g., "Guns kill." However, is technology truly independent of society? It is often countered that society determines technology, e.g., "Guns don't kill, people do." The problem with this back-and-forth argument is that we soon run into an impasse. STS offers a way out by re-conceptualizing the relationship between society and technology. Are they independent or mutually exclusive? No. Technology is social through and through. Even though artifacts are made of different stuff than you and I, they are of the social world. Design and implementation are affected by social, economic, and political decisions. This is true not only for the most controversial technologies, such as nuclear reactors, surveillance cameras, and encryption software but also for bicycles (see Bijker 1995) and air bags (see Wetmore 2004).

Conversely, society is technological through and through. In his wonderful essay, "Where Are the Missing Masses?" Bruno Latour (1992) notes that many artifacts are crucial to social order. They close doors (pneumatic hinge), promote traffic flow (stop signs and lights) and serve seatbelt legislation (caution chimes or lights), for example. Alter them and you change society. When the electricity grid fails, those affected lose illumination, refrigeration, air conditioning (or heat), mass entertainment, communication mediums, and computer access. Imagine if the grid stayed down for years. Wouldn't we lose modernity?

Eliminating the false dichotomy between society and technology, we can consider ways in which sociotechnical ensembles develop. This allows for a more constructive approach to seemingly intractable social dilemmas. For instance, we would assert that the problem is not guns per se, nor violent people per se, but a gun-violence ensemble (a volatile mixture of inexpensive handguns, lax background check system, poverty, gang culture, bullying, etc.) and look for systematic correctives. We can also better appreciate sociotechnical controversies like the one we are investigating when we see that the transhumanity debate is not simply over allegedly good or evil technologies and, instead, it is about how technologies function in a particular social order.

Discourse of Concern and Discourse of Hope

Nancy Campbell defines "suspect technologies" as those perceived with "reasonable suspicion that their development, deployment, and effects are unevenly distributed, differential, and more likely to be socially unjust than not" (2005:375). Using Wiebe Bijker and Trevor Pinch's terminology, we could consider "suspect sociotechnical ensembles" when suspicions encompass technologies *and* social institutions. With regard to neuropharmacology, for example, Ad Bergsma (2000) warns of pharmaceutical companies manipulating consumers by marketing designer drugs that tinker with the brain's pleasure center. Francis Fukuyama (2002) predicts that if genetic engineering of children is allowed the most powerful families will surely extend their advantages, thereby perfecting aristocracy. One of Frodeman's worries regarding nanotechnology is that extremely small surveillance devices placed within one's body would be used by authorities as social control mechanisms.

When Jasper Lassen and Andrew Jamison (2006) asked focus group members to talk about genetic technologies (agricultural and medical applications) they found a very rich "discourse of concern." The respondents raised questions not only about technologies but also about whether agribusiness and state regulators could be trusted. They raised philosophical questions about the natural order. In my own research with college students I found that young men and women critically assess biotechnology in terms of their understanding of God's plan.

At times it can be bewildering following the strands of points and counterpoints woven throughout the transhumanity debate. The discussion is over social equality and then it moves to the nature of consciousness, the rate of technological innovation, etc. The debate is sprawling! Even so, one can identify the arrangement of a "discourse of hope" pitted against a "discourse of concern." Transhumanists generally appreciate modern society, economy, science, and technology and they imagine that social life will only improve as the ensemble matures and its full potential is realized. Simon Young (2006) refers to this as the "New Enlightenment." Their adversaries disagree in one of two ways. Conservative critics voice misgivings regarding modernity and fear that transhumanity will make matters much worse. Liberal critics worry that pushing forward with transtechnologies will destabilize the modern ensemble, thereby erasing gains made over the past few centuries.

Transhumanity and Modernity

For the most part, contestants are not interested in making distinctions between modern and postmodern developments in culture, society, and economy. In keeping with the debate, for our purposes modernity will stand for the contemporary state of national and global systems. Conservationists disagree about the relationship between modernity and transhumanity with critics of modernity seeing a sequel and advocates of modernity perceiving a revolution.

Suspect Modernity

In the Foreword to *Brave New World*, Aldous Huxley asserts that "It is only by means of the sciences of life that the quality of life can be radically changed…This really revolutionary revolution is to be achieved, not in the external world, but in the souls and flesh of human beings" [(1932) 1969]. He imagines a totalitarian government utilizing technologies for social control purposes. The "World Society" in his dystopia involves embryonic and fetal biochemical engineering, neo-Pavlovian conditioning, hypnopædia or sleep-learning, and powerful psychotropics to mass produce members incapable of contesting their caste standing and the social system. Whereas Max Weber warned about bureaucracy and an "iron cage," Huxley raises an alarm about science and a bioengineered cage.

Leon Kass, a noted ethicist and a former Chairman of the President's Council on Bioethics, explicitly draws many parallels between Brave New World technologies and current ones:

> Our Prozac is not yet Huxley's "soma"; cloning by nuclear transfer or splitting embryos is not exactly "Bokanovskification"; MTV and virtual reality parlors are not quite the "feelies"; and our current safe and consequenceless sexual practices are not universally as empty or loveless as those in the novel. But the likeness between Huxley's fictional world and ours are disquieting, especially since our technologies of bio-psycho-engineering are still in their infancy, yet vividly reveal what they may look like in their full maturity (2002:5–6).

As with Huxley's World Society, Kass believes that consumer society promotes a "soft dehumanization" and, if allowed, capitalist enterprises will continue to use science and engineering to provide every manner of indulgence in exchange for dependency: "Homogenization, mediocrity, pacification, drug-induced contentment, debasement of tastes, souls without loving and longings—these are the inevitable results of making the essence of human nature the last project for technical mastery" (2002:48). Carl Elliott warns that the commercialization of transtechnologies amounts to unprecedented market access to the body and mind:

> [F]or critics of genetic enhancement, the market represents something far more sinister because it seems to view the world as a place where everything has a price. How will our sensibilities be changed if we start to see our children, our bodies, and our minds as potential objects of consumption? Where does the soul go, once it's been priced and tagged? (2003).

Bill McKibben (2003) asserts that the intense competition of the capitalist labor market will drive reluctant parents and individuals to risk the side-effects of increasingly potent enhancers simply to compete effectively over academic placement and good jobs. He worries that satisfaction and pride in one's accomplishments will be compromised.

Modernity in the Balance

Jurgen Habermas (2003) and Francis Fukuyama (2002) value the liberal state and are mainly concerned that transhumanity may undermine it. Habermas asserts that democracy works as long as citizens accept the right of fellow citizens to participate. This, in turn, is based on citizens' assumption that the other men and women going to the polls and serving in office are essentially the same as they in terms of their capacities. As it stands now, children, chimpanzees, dolphins, and dogs are not allowed to have a say in the governance of society, in part, because they are thought not to have the same capacity for reasoning. He alleges that if transhumans, and then posthumans, become cognitively or emotionally distinct from humans, commonality will cease and the polis will fracture. Habermas also warns that genetically-designed children will not be perceived as autonomous actors because they will be governed by the "irreversible intentions of third parties" (63). Once again, at stake is universal egalitarianism.

For Francis Fukuyama liberal democracy is a crowning achievement. His bold declaration of the "end of history" was meant to signify that we have left the rough and tumble period of failed political and economic experiments and have reached the point of fine tuning. However, he warns that all this could change. Religious fundamentalism and terrorism represent serious threats to democracies, but Fukuyama is more concerned about human life technologies. What if human nature is reconfigured such that citizens become more docile, more manageable? Authoritarian government would be so much more formidable with a population incapable of dissent. Fukuyama warns that transhumanity will open a door for a return of totalitarianism.

The "New Enlightenment"

The transhumanists respond to their critics by chastising them for a lack of faith in the vitality and durability of the Western political economic system. They certainly do not want to see a resurgence of totalitarian regimes, and they assert that as long as we stand firm and not allow a dismantling of citizens' rights we should avoid that fate. They trust that free citizens will not opt for technologies that limit self-determination.

The true villain in Huxley's dystopia, according to the transhumanists, is state power. Give the World Society or any government too much power over its citizens and there will be grave consequences. Any technology deployed by a totalitarian government for social control purposes will be harmful. You prevent this by restricting the degree to which the state can dictate personal matters and by protecting individual rights, especially the right to make decisions regarding one's own body and relationships.

Liberal political philosophy is enshrined in constitutions and it is used to justify, among other things, free speech, artistic expression, parental rights, and reproductive choice. Human rights are often expressed in terms of freedoms that all persons are entitled to enjoy. They can be stated as freedom *from* coercions, for example, not to be held in slavery or suffer torture, or freedom *to* act, for example, to practice religion or express opinions. Ratified in law, they frame a protected space for personal autonomy and self-determination. From the United Nations' Universal Declaration of Human Rights (1948) we have these: "the right to life, liberty and security of person" (Article 3), freedom from "arbitrary interference with his [sic]privacy, family" (Article 12), "freedom of thought, conscience and religion" (Article 18), and freedom "to enjoy the arts and to share in scientific advancement and its benefits" (Article 27). The universality of these rights, i.e., that they be accorded to all people regardless of gender, race, property ownership, beliefs, and creed, satisfies a standard of morality as expressed in the Golden Rule and Immanuel Kant's categorical imperative, which in this instance might be stated as "the liberties that we most desire for ourselves should be provided to all." Furthermore, as John Locke argued, these rights must be provided to *all* if tyranny and discrimination are to be restrained.

Although James Hughes, a leading figure in the World Transhumanist Association, may use the postmodern-sounding "citizen cyborg" in the title of his 2004 book, he makes it clear that this concept of citizenship is based on the political philosophy of John Stuart Mill, whom he quotes at least five times in his text. He praises Mill's rationale for liberty and calls for an extension of liberal democracy. For example, the right to control one's own body should be extended to include the right to augment one's body. Enfranchisement should be extended to transhumans or posthumans. As Hughes sees it,

> Transhumanism is a direct product of this radical democratic tradition. Transhumanists, like their democratic humanist forebears, want to create a global society in which all persons, on the basis of their capacity for thought and feeling, can participate as equal

citizens, control their own affairs and achieve their fullest potential, regardless of the characteristics of their bodies (2004:81–82).

A classic question regarding governance can be posed in terms of technology: Who should we trust to decide our technological future, an elite or all citizens? Gregory Stock (2002) answers that, despite trepidations, it is best to trust the people. Autocratic approaches produce terrible results due to heavy handedness, inflexibility, and because collective wisdom is spurned. Stock believes that most people act prudently. If certain applications appear too dangerous or repulsive, consumers will not support them and they will fail in the market. Parents, more so than anyone else, look after their children, and will only embrace safe and efficacious technologies for their daughters and sons. According to Stock, "[t]echnology doesn't emerge magically; it depends on the existence of large numbers of people who want it. Today we are actively choosing the technologies that serve us, and if future generations do the same, people's biggest fears will not come to pass" (151).

James Hughes dismisses the assertion that parental say over the design of a child's genome will be detrimental. Such genetic decisions should be understood along the lines of parents' provision of nutrition and education for the overall development of the child. Rather than being fetters, genetic modifications are more likely to serve creativity:

> Few parents intend to make their children less intelligent or less capable of autonomy and communication. If anything, parents' choices will generally expand children's ability to communicate, make decisions, and control their own lives...[i]f there were widespread evidence that parents were systematically choosing to make their children less capable of making choices, less able-bodied, less intelligent, then I would be for regulating those bizarre choices (2004:149).

He concedes that some families will choose enhancements while others will not, and over time the population will diversify. He recommends that national and international law be amended, if necessary, so that political rights are fairly and uniformly provided for *all* intelligent persons, human or transhuman. Just as the polis did not fracture with the incorporation of women and liberated slaves, he contends that with the proper legal framework in place the political community can handle diversity of another kind.

In concluding this section, I'd like to point out that if we accept the accusations leveled by both sides in the debate we would be in the untenable position of "dammed if you do, dammed if you don't." The conservationists contend that if liberal democracies go down the transhumanist path and allow free choice for enhancement technologies, consumers knowingly or unknowingly will suffer modifications that diminish free will. Governments would exploit them. According to the transhumanists, if liberal democracies take up the conservationists' cause, the state would become more involved in the regulation and control of reproduction, the body, and parenting through banning enhancement technologies, and monitoring and policing illicit use. This would entail an increase in state power and loss of personal autonomy.

Each side raises the specter of totalitarianism. They dispute which truly is the destructive path, but not the high stakes involved. However, as I will discuss in Chap. 4, there is a tendency in debate rhetoric to overemphasize and we should not accept without critical review the claim that the fate of democracies hang in the balance.

On Capitalism

Whereas most conservationists disfavor the increasing interconnection between science and business and register concerns about the direction of global capitalism, transhumanists favor entrepreneurial capitalism. Many have firsthand experience with hi-tech ventures and feel quite comfortable with what they see as the mutually beneficial relationship between science, engineering, and business. As long as the system continues to support innovation there is little reason to oppose it. Why kill the goose that lays the golden eggs?

We should, however, be careful not to overstate their position towards the free market system. There is a tendency to mistake the transhumanists' unequivocal admiration of capitalist productivity for economic libertarianism, i.e., the belief that the free market system should be left to operate unencumbered by the state. An economy provides for the production of goods and services in a society, but it also functions to allocate goods and services. In a capitalist economy, allocation follows effective demand, in other words those that have the means to pay for what they want, get what they want, and those without the means, go without. Transhumanists hold a very mainstream view that certain goods and services are so important to the greater good that the state should intervene to ensure universal allocation. Such public goods now include education and immunization, and the transhumanists want future life-enhancing technologies to be included as well. They offer this in the spirit of social welfare, but they also worry that social conflict might arise if only the affluent have access to them.

Conclusion

It may come as a surprise that transhumanists are very supportive of the modern social system. A mixed economy is fine. The state has a role to play in promoting social welfare but it should not be allowed to dictate personal matters. Democracy must prevail over fundamentalism. At least in terms of political-economic leanings, transhumanism is not a radical ideology, not even a reform ideology. We need to keep in mind that all ideologies, including transhumanism, are designed to serve movement interests. Transhumanists keep their eyes on the prize. If there is every expectation that biotechnology, nanotechnology, neurotechnology, and computer technology will continue to flourish under the political economy of Western societies and the global system, it makes perfect sense to back these.

In the following chapters we'll have the chance to explore in much greater detail the transhumanist argument and the conservation counterargument. We'll take up the dispute over whether transhumanity will amount to transcendence or transgression. What is the implication for human body and mind? What are the risks? Is transhumanity inevitable? However, I'd like to conclude this chapter by modifying my initial characterization of transhumanism as a dangerous ideology. The call for human perfectibility is certainly audacious but transhumanism will not go far if it is too shocking and unconventional. The fact of the matter is that transhumanity is put forward wrapped in the familiar package of Western progress. Transhumanists envision a future of robust democracy, compassionate government, and prosperity. Such optimism is uncommon today and may invite curiosity and interest in their more controversial ideas. Their opponents reject the core element of transhumanity which is human enhancement but they must also cast doubt on the transhumanists' vision for the future. As we will see in the next chapter this is done by countering transhumanists' call for engineered transcendence with dire warnings of mankind's fall.

Chapter 2
Transcend or Transgress?

Certain passages from C. Wright Mills' *The Sociological Imagination* [(1959) 1969] inspire me today as they did when I happened upon my first sociology course many years ago. In the first chapter, *The Promise*, Mills identifies three sets of questions posed by the classic social theorists such as Karl Marx and Max Weber. It is the third set, especially, that I find most insightful:

> What varieties of men and women now prevail in this society and in this period? And what varieties are coming to prevail? In what ways are they selected and formed, liberated and repressed, made sensitive and blunted? What kinds of 'human nature' are revealed in the conduct and character we observe in this society in this period? And what is the meaning for 'human nature' of each and every feature of the society we are examining?

Underlying these questions are three premises:

- Subject and social world (or biography and history, as Mills puts it) are interrelated.
- Human nature is flexible.
- What prevails today, systems and subjects, will not prevail in the future.

Karl Marx studied the capitalist system with concern for worker alienation and class conflict. Max Weber wrote about rational management and social control. The contestants in the transhumanity debate are from diverse backgrounds and on a number of issues they do not see eye to eye, nevertheless, like Marx and Weber all accept the premise that the human subject is influenced by social/technological forces. It is obvious that the transhumanists see human nature as changeable, but the conservationists' call for preserving it belies a similar attitude. Arm-chair philosophers? Hardly. They, too, are activists and are engaged in this debate because for them the future of humanity is at stake.

In an article published in *Christianity Today*, entitled "The Techno Sapiens Are Coming," C. Christopher Hook (2004:36) begins by warning, "When God fashioned man and woman, he called his creation very good. Transhumanists say that, by manipulating our bodies with microscopic tools, we can do better. Are we ready for the great debate?" After identifying a few transhumanists and quoting their

S. Lilley, *Transhumanism and Society*, SpringerBriefs in Philosophy,
DOI: 10.1007/978-94-007-4981-8_2, © The Author(s) 2013

more dramatic lines (such as "biology is not destiny…chips are destiny" and "the age of the human is drawing to a close"), he asks and answers the rhetorical question, "Are these ideas the musings of a small band of harmless techno geeks? Unfortunately not."

Leon Kass describes transhumanists as a social movement vanguard :

> In leading laboratories, academic and industrial, new creators are confidently amassing their powers and quietly honing their skills, while on the street their evangelists are zealously prophesying a posthuman future. For anyone who cares about preserving our humanity, the time has come to pay attention (2002:4).

We should question this particular rhetorical strategy and not assume commensurability between researchers and transhumanists. For the most part scientists and engineers are involved with what Thomas Kuhn (1962) calls "normal science," that is, contributing incrementally to established lines of research. They are busy with the day-in and day-out routines of administering projects, running labs, and securing grants. Funding is more readily available for research related to the detection, understanding, and treatment of pathologies and, understandably, researchers often present their work and findings in terms of potential therapies (not enhancements). With some notable exceptions, for instance James Watson, most are not flamboyant and are far too prudent to make political waves.

Furthermore, although it is tempting to cast scientists and engineers as Dr. Frankensteins, some researchers have found that experts, as compared to non-experts, are not more likely to throw caution to the wind. For example, Isaac Rabino (2003) found that human genetics researchers had similar attitudes about genetic testing as that of the general public with regard to supporting paternalism when dealing with the test results of children, favoring voluntary testing over compulsory testing, and opposing disclosure to insurers and employers. Lennart Sjöberg (2002) found that experts and non-experts, alike, worry about tampering with nature and novel risks.

Whether or not scientists and engineers favor engineered transcendence is debatable, but we know for sure that the transhumanists *explicitly* propose it. They are the visionaries. This is evident right from the start with Julian Huxley's coining of the term transhumanism:

> The human species can, if it wishes, transcend itself—not just sporadically, an individual here in one way, an individual there in another way, but in its entirety, as humanity. We need a name for this new belief. Perhaps *transhumanism* will serve: man remaining man, but transcending himself, by realizing new possibilities of and for his human nature (1957:17).

Transcendence: Cosmic, Personal and *Civitas*

Diversity exists in any movement so it should come as no surprise that there are different versions of transcendence espoused by transhumanists. I will present three versions: cosmic, personal, and *civitas*. I will present each according to its expression by a prominent figure in the transhumanist movement, however I want to make clear that with selective sampling and refinement I am actually proposing ideal types. Following that, I will present the conservationists' rebuttal to claims of transcendence: that radical bio-social change will bring about fatal transgression.

Cosmic Transcendence

Of the bizarre states that cosmologists describe, the singularity is striking. It is posited to be a feature of a black hole, which itself is a *very* strange phenomenon. A black hole is born when a collapsed star forms a region of space with an extraordinarily intense gravitational field. Under such conditions, all matter flows through a single point, or singularity. The initial state of the universe, prior to the Big Bang, is also described as a singularity. Laws of time and space do not operate within a singularity. Ray Kurzweil used this concept as a key metaphor and title for his 2005 book, *The Singularity Is Near: When Humans Transcend Biology*, to characterize a point in the future when socio-technological change will be beyond anything we have ever known and standard theories of human development will fail. He predicts that breakthroughs in one cutting-edge field, for example, neuroscience, genetics, robotics, and computer science, will spur innovations in the other fields. Convergence will help produce exponential growth in the *rate of change* that will make Alvin Toffler's future shock look like a stroll in the park.

Physicists explain that matter passing through the singularity of a black hole is dramatically affected. In Kurzweil's model, as humans pass through the technological singularity an accelerating evolutionary process working on intelligence will yield new beings. He makes it clear that there will be as little in common between posthumans with evolved intelligence and standard humans as there is between bacteria and Homo sapiens. He predicts that these super beings will harness stars and eventually operate on the scale of the universe or universes. Generations of humans that forego this evolution, in comparison, will be hopelessly primitive.

In the social sciences it has been pointed out that grand theories, such as Herbert Spencer's social evolution theory, lose sight of human actors. Kurzweil's evolutionary theory is even more sweeping than Spencer's, suggesting stages of civilization freed of bodily *and* earthly constraints. With cosmic transcendence, human actors recede as anachronisms.

Personal Transcendence

Max More, in promoting an actor-oriented approach to transhumanity, has openly expressed impatience with epic scales, and has targeted the singularity for scorn:

> The Singularity idea has worried me for years—it's a classic religious, Christian-style, end-of-the-world concept that appeals to peoples in Western cultures deeply. It's also mostly nonsense…The Singularity concept has all the earmarks of an idea that can lead to cultishness, and passivity. There's a tremendous amount of hard work to be done, and intellectually masturbating about a supposed Singularity is not going to get us anywhere (quoted in Hughes 2004:173).

For more, transcendence is primarily a personal experience, a process of self-transformation. This is best expressed in his *Principles of Extropy* (2003) which he crafted while serving as chairman of the Extropy Institute. In defiance of entropy as experienced by individuals as disease and decline, he recommends the bold application of enhancement technologies for *extropy*: "seeking more intelligence, wisdom, and effectiveness, an open-ended lifespan." (2003) Essential to transcendence is one's will to advance and one's intolerance for passivity. One must embrace rational thinking over faith that constrains and one should challenge traditional notions of human limitations. Believing in perpetual progress and being proactive and optimistic vis-à-vis science and technology leads one "creatively and courageously to transcend "natural" but harmful, confining qualities derived from our biological heritage, culture, and environment." (2003) He values an open society through which individuals may self-direct and voluntarily cooperate to secure advantages.

More [(2000) 2006] finds in Friedrich Nietzsche's overman a prototype for transhumans. He quotes this passage from *Zarathustra* II: "And life itself confided this secret to me: "Behold," it said, "I am *that which must always overcome itself.* Indeed, you call it a will to procreate or a drive to an end, to something higher, farther, more manifold: but all this is one…Rather would I perish than foreswear this…"

In calls for moratoriums or for the relinquishment of advanced technologies, More sees stagnation. In the bold self-application of enhancement technologies he sees, instead, a great opportunity to engage the creative spirit of the overman.

Civitas Transcendence

James Hughes, a former secretary of the World Transhumanist Association, offers a vision of transhumanity in his book, *Citizen Cyborg* (2004), which is meant to be consistent with secular humanism and the Enlightenment project of using science and technology for the collective good. His training as a sociologist shows through with his attention to social and political systems. He advocates improvements to minimize social injustice, promote social solidarity, and safeguard human populations. Like Max More, he finds cosmic transcendence so abstract and future distant to be a distraction for immediate concerns, but he also finds fault with the libertarian streak of the extropians. Hughes distrusts the free market, opposes unchecked individualism, and believes that a safe passage to a transhuman civilization requires ethical standards, public oversight, and some regulation.

I use the Latin term, *civitas,* which denotes citizenship and also planned settlement, to describe Hughes' vision of transcendence. He foresees the progression to a more just, equitable, prosperous, and peaceful world through democracies that encourage citizens to utilize safe and effective enhancements. Because they are augmented by biotech, nanotech, and neurotech, *cyborg citizens* will be more capable and energetic citizens and be able to contribute more to community and society. A virtuous spiral develops such that as enhanced citizens become more socially productive, societal goods increase, as more individuals share in this bounty, their quality of life increases and, in turn, they contribute more to the common good.

As a way to promote egalitarianism, Hughes proposes social welfare programs designed to assist those who can't afford enhancements. He also agrees with a policy recommendation by Nick Bostrom (2005) that "positional enhancements" that benefit an individual at the expense of others should be discouraged or banned. Hughes is quite aware of how the counter tendencies of selfishness/altruism, self-centeredness/empathy, and conflict/cooperation can be influenced by social institutions, social groups and culture. However, he also entertains the possibility that these tendencies are rooted in biology. He favors Mark Walker's suggestion that more research be conducted "identifying the genes and neurochemical necessary for empathy and cooperation, encouraging noncoercive screening and therapy to ensure that all citizens have them, and giving incentives for people to select for them in children and amplify them in themselves" (251).

To the extent that this policy is aimed at shaping the human population, he is recommending a form of eugenics. Hughes, however, distinguishes this policy from discredited totalitarian practices in that it is voluntary and is not motivated by prejudice. It is meant to achieve a greater good, namely, to develop cyborg citizens better suited for democracy. Whereas Kurzweil values science and technologies for the lift that they might provide for superior intelligences, and More values these as resources for the overman, I see Hughes following Saint-Simon and Auguste Comte and embracing science and technology for the purpose of social engineering. Granted, he would not have this done in a heavy handed way and he defers to "cognitive liberty," nevertheless he imagines a transhuman future in which "pro-social feelings" are a requirement for public service employment and all are under an "ethical obligation... to enhance ourselves, to become better people and use our powers to do good" (256).

Compromise between Versions

In my account above, I note contested points between the three models of transcendence. However, it is possible to reduce tension if certain accommodations are made. For example, those attracted by personal transcendence could imagine cosmic transcendence taking place at time well after an initial phase of recognizable self-transformation. The sticking point is over the expected or desirable rate of change.

The libertarian streak in personal transcendence can be muted a bit by social pragmatism. Hughes makes the case that without public support transhuman technologies will be delayed or banned, and the public will accept enhancement technology only if it is safe, broadly available, and democratically accountable. Seen this way, self-interest in transcendence will depend on accommodating collective goals. Max More (2005) appears to concede this point in a policy position paper in which he recommends a "proactionary principle" that retains the freedom to innovate but adds, for example, openness/transparency: "Take into account the interests of all potentially affected parties, and keep the process open to input from those parties."

Transgression

Regardless of the version espoused, transhumanists envision a progressive series of technological innovations and enhancements with every new stage of transhumanity being better than the last. In contrast, conservationists warn of transgression, or a point of no return from which humanity will suffer a most grievous, irretrievable loss. Although conservationists may not make a distinction between the three versions of transcendence, I have distilled the respective critiques and present them below.

Critique of Cosmic Transcendence

Max More is not alone in his accusation that Kurzweil's singularity is a high-tech version of Christian eschatology. Conservationists also see in cosmic transcendence a quasi-religious theme that contradicts the ostensibly secular, scientific basis of Kurzweil's work. To explain the comparison, it might help to contrast Pierre Teilhard de Chardin's vision of the Omega Point with more militant versions of Christian eschatology.

John, in the Book of Revelation (New Testament), reports a vision of divine intervention and judgment, with Jesus Christ returning to earth to save the righteous and vanquish satanic forces. Countless times throughout the centuries, Christian groups have anticipated Armageddon. More recently, Jerry Jenkins and Tim LaHaye, have popularized this scenario with their *Left Behind* series, which by 2008 had sales surpassing 65 million copies. Pierre Teilhard de Chardin's vision is very different. He believed that all elements of the universe are imbued with spirit and are involved in an evolution of consciousness. Human beings represent an important stage in this development because of their self-consciousness. He expected ever higher levels of consciousness will emerge out of increasingly complex human-world interactions and eventually the Omega Point of supreme consciousness will be reached. Far different from the anthropomorphic warrior and judge of Revelation, Jesus Christ/God is imagined by Teilhard as an elemental, universal force behind the evolution of consciousness, drawing humanity closer through ever-more complex manifestations.

There are significant similarities between Kurzweil's vision of the future and Teilhard's eschatology including an emphasis on consciousness, an evolutionary theory that provides for superorganisms (emergence of complexity from more basic forms), an exceptional role for humanity but also its superannuation, and the culmination in universe consciousness. The differences, however, should also be noted. Pierre Teilhard de Chardin describes consciousness in terms of spirit whereas Ray Kurzweil emphasizes information, computation, and intelligence. He and other transhumanists describe the evolutionary process in secular terms, for example, natural selection, Moore's Law of exponential increases in processing power, and technological innovation, so that there is no need to introduce divine agency.

Kurzweil doesn't consider singularitarianism to be a religion, although he acknowledges that it provides "new perspectives on the issues that traditional religions have attempted to address: the nature of mortality and immortality, the purpose of our lives, and intelligence in the universe" (2005: 370). He speculates about the divine with respect to the saturation of the universe with intelligence, but he treats it as an open question whether posthumans with god-like powers will find an eternal God.

Critics operating from faith traditions charge that Kurzweil's model of cosmic transcendence lacks an accounting of God's involvement with individuals and humankind. They take exception to the elevation of science and technology as the agents of history. Michael DeLashmutt (2006) writes: "Though a posthuman eschatology wrestles with similar themes present within Christian eschatology, a Christian eschatology is ever aware that the fulfillment of its hope lies in the hands of the God who is in control of history, in contrast to a posthuman eschatology that places the onus of control upon human technologies." In the encyclical, *Spe Salvi,* Benedict XVI asserts that

Francis Bacon and those who followed in the intellectual current of modernity that he inspired were wrong to believe that man would be redeemed through science. Such an expectation asks too much of science; this kind of hope is deceptive. Science can contribute greatly to making the world and mankind more human. Yet it can also destroy mankind and the world unless it is steered by forces that lie outside it... It is not science that redeems man: man is redeemed by love...If this absolute love exists, with its absolute certainty, then—only then—is man "redeemed", whatever should happen to him in his particular circumstances (2006:26–27).

Elaine Graham (2003) warns of "hyper-humanism":

Such talk of humanity as in some degree self-constituting via its own technologies, of being capable of influencing the course of its own development is to fall prey to what we might term 'hyper-humanism': a distortion of modernity's faith in the benevolence of human reason, producing the hubristic belief that humanity alone is in control of history (2006).

She believes that humility before God and creation is especially necessary in the near future as more and more powerful technologies become available. With humility comes caution, reflection, and prudence. This disposition may prevent catastrophes. Bronislaw Szerszynski (2006) fears that technologies misconstrued as angels for humanity may become demonic: "the irony is that the denial that technologies belong to God seems ultimately to give them not to us, but to *themselves*—to render them demonic, and to place humanity under their thrall." Alan Padgett (2005) is pessimistic as well:

> The dream of a happy and harmonious techno-secular future is based on false hopes in infinite energy, infinite human potential, infinite human progress, and complete human good will. Such a techno-secular dream, even if it comes about, will self-destruct after a few centuries, inevitably smashing on the rocks of our finitude and sin.

Transgression is imagined within a Judeo-Christian tradition that, as with the stories of Adam & Eve's original sin, the Golden Calf, and Tower of Babel, warns not to put humanity or its creations above God. Pride goes before destruction (Proverbs 16:18).

Whereas these critics see problems with transhumanism being insufficiently attuned to divine grace and God's plan, secular critics find fault with it for being too influenced by Christian eschatology. David Noble, in particular, has advanced the thesis that Western science and technology were inspired by Christian millennialism and these institutions remain essentially religious endeavors directed today by men motivated by a quest for transcendence. According to Noble, the intellectual movement began in Europe in the Middle Ages among monastic orders such as the Benedictans and Franciscans with Erigena, Roger Bacon, and Francis Bacon, among others, calling for the development of technologies to better achieve these religious goals: (1) to recover the powers of dominion that had been lost with the Fall of Adam, (2) to allow man to better appreciate his likeness to God the Creator, and (3) to use the powers to wage a successful campaign (deemed imminent) against the Antichrist and his forces. Noble asserts that Newton, Boyle, Priestly, Faraday, Maxwell, Babbage and many other notable scientists and technologists were believers and, in the nineteenth and twentieth centuries, the fields of nuclear physics, space exploration, artificial intelligence, artificial life, and genetics were launched by men inspired by Christian eschatology. Allegedly, religious and non-religious scientists and engineers in these fields today continue to be obsessed with the quest for perfection: "Often displaying a pathological dissatisfaction with, and deprecation of, the human condition, they are taking flight from the world, pointing us away from the earth, the flesh, the familiar" (1999:208).

In summary, critics of cosmic transcendence may disagree whether the roots are Christian or humanist, nevertheless they find fault with its techno-utopianism and its impatience for human limitations. Noble's recommendation: "disabuse ourselves of the other-worldly dreams that lie at the heart of our technological enterprise, in order to begin to redirect our astonishing capabilities toward more worldly and humane ends" (6).

Critique of Personal Transcendence

Critics often accuse transhumanism of promoting excessive individualism. This charge, however, somewhat misses the mark with cosmic transcendence, nor does it apply to social-political transcendence with its communitarian leanings. The apt target is personal transcendence.

As described previously, More's account of personal transcendence is fashioned after Nietzsche's overman, but of course there are other possible models, for example, the "self-made" entrepreneur and the conquering heroes of antiquity such as Alexander the Great and Julius Caesar. As with these other models, personal transcendence makes the self the overriding project of one's existence and as such it requires a significant preoccupation—how to best utilize resources, how to maximize potential, etc.

According to religious critics, as self-absorption increases there is a corresponding decrease in the tendency to enter into rich reciprocal relationships with others, especially if self-sacrifice is required. There is also a tendency to treat one's environment in terms of use value. What if God is encountered through a reverence of creation and through "I-Thou" relationships (Martin Buber's term), as many theologians assert? What if self-abnegation is necessary to open oneself to the divine? William Schweiker (2003) asserts that "The daring task about speaking about the divine aims to articulate realms of value beyond human preference and power. And it seeks also to evoke a love of life rooted in the reality of the living God."

For many believers in the Abrahamic faiths, human nature is God-given (according to Genesis), passed down securely through generations, and designed for a higher purpose, for example to be endured as a pre-condition for God's grace and redemption or, more optimistically, embraced– bearing the likeness of God allows for a meaningful relationship with the divine. John Jefferson Davis asserts that

> All of God's creation, including the human body, is *good* (Gen. 1:31; Tim. 4:4) and as such is worthy of care and respect. Human beings occupy a unique place in creation, being made in the image and likeness of God (Gen. 1:26), and consequently human life has sacred value and is to be accorded the greatest care and protection…(70).

In this light, enhancements pose a serious threat. Mark Hanson (1999) writes that "[w]ithin a Protestant understanding of our nature, the disvalue occasioned by enhancements might consist… in the loss of recognition of the providence of God working through the contingencies and weaknesses of our human form."

There is a sense of daring and excitement associated with personal transcendence. The self is emboldened, tested, and augmented through enhancements. Those who question this process must advise, instead, self-restraint and self-sacrifice. I imagine that this is not an easy task. Individuals must be persuaded to believe in an apparent paradox, that something is gained through selflessness and something is lost through self-fulfillment. It requires, most of all, relying on a model of character development that Leon Kass, for one, admits is a bit old school.

The four cardinal virtues from Greek philosophy are prudence, temperance, fortitude, and justice. Islam recognizes those, as well as, righteousness, respect, sincerity, and honesty. Christianity adds faith, hope, charity, and love. Buddhism's Divine States are loving kindness, compassion, altruistic joy, and equanimity. Practicing these virtues requires self-restraint and generosity towards others. Vices, for example, pride, avarice, and gluttony are typically described as manifestations of selfishness. Although the following is a very simplified formula, excellence of character or proper living is said to be achieved through practicing virtue (which is self-effacing) and avoiding vice (selfishness).

Is personal transcendence consistent with this formula? "No," assert the critics of transhumanity. It is egotistical, too grasping, and may result in new forms of injustice. Living a good life accepting of human mortality, on the other hand, has intrinsic value and it helps promote the greater good. Worried about overpopulation that may occur with elongated life spans and increased demands placed on natural systems, Bill McKibben sees finite living as the choice consistent with conservationism.

Transhumanists treat death and decline as major impediments to overcome. Simon Young (2006) bluntly states, "Death is, to me, an obscenity" (15) and he refers to illness, disability, and senescence as "biological slavery" (41) One's existence, in his view, takes place only within life's frame. There is no afterlife. Conquering death is a way to extend life's frame. In terms of the overman, moreover, the will is strengthened through death's conquest.

Clearly, this understanding of and approach to death is at odds with that of most religions. Rather than treat it as the tragic end of the person, death is understood as a passageway to a better state of existence—the absence of suffering, peaceful co-existence with others, a more perfect union with the divine. John Paul II (1981) relates suffering and death in terms of Jesus' crucifixion. By accepting these, as did Jesus, we also "carry the cross." This is an act of homage as well as a vital exercise in one's spiritual development.

Bill McKibben explicitly rejects the transhumanist premise that increased longevity is a necessary condition for self-fulfillment. He argues that the standard human lifespan is sufficient time to lead a fulfilling life, and that finitude actually makes life more precious. Life extension and enhancements will dilute human experience and undermine character development as individuals will dodge true adversity. Personal accomplishments will ring hollow for the enhanced. According to McKibben, the "grander questions" regarding human existence "can only be usefully answered by people, whose bodies eventually start to sag, by people who love and who grieve and who celebrate, by people who mourn and who know that they will someday die" (2003:226).

Critique of Civitas Transcendence

James Hughes charges that "Left bioLuddites" have turned away from their roots in the Enlightenment: "They have given up on the idea of progress guided by human reason, and, afraid of the radical choices and diversity of a transhuman future, are reasserting mystical theories of natural law and order" (2004: xiii). Conservationists counter that engineered transcendence will backfire and undermine the humanist project.

Martin Heidegger [(1954) 2003] alleged that with modern technology there is a particular revealing and ordering of being that treats people as a "standing-reserve" to be exploited. In his *Case against Perfection*, Michael Sandel warns that enhancement engineering also entails a disposition of mastery that works against the Enlightenment ideal of liberty: "willfulness over giftedness, of dominion over reverence, of molding over beholding" (2007:85). The social theorist, Jurgen Habermas (2003), is particularly concerned with preimplantation genetic diagnosis (PGD) and biotechnologies that allow for the selection or modification of a child's genes. Habermas warns that a designed child will not be an autonomous agent and will not be perceived as such.

Can a child have true autonomy if parents genetically design his or her capacities and proclivities? Maureen Junker-Kenney believes the answer is no:

> Genetic enhancement exemplifies a total reversal of the preconditions for autonomy: The offer of pre-implantation enhancement and selection constitutes the victory of parents' projections over the otherness of the child. In co-creating the specificies of its reality—sex, bodily features, character predispositions—it is being denied the singularity that is based on an unmanipulated originality (2005:12).

She asserts that the parent that designs his or her child would gain unprecedented influence over the child. Habermas writes that from the child's perspective, this is "permanent dependence" and "[f]or this poor soul there are only two alternatives, fatalism and resentment" (2003:14).

The questionable legal and moral standing of designed humans raises this dilemma for the polis: If granted citizenship these individuals may not be trusted as jurists, voters, and public officials but if denied political rights the promise of inclusion will be denied. Modern states operate pluralistically by recognizing human commonality. Although Hughes believes that this system can accommodate transhumans, Habermas insists that it cannot.

Francis Fukuyama claims that time and time again regimes have attempted to control subjects through systematic social control mechanisms only to be thwarted in the long run by unruly human nature. He asserts that there are "natural desires, purposes, traits, and behaviors [that] fit together into a human whole" (2002:12) and that these "deeply rooted natural instincts and patterns of behavior reassert themselves to undermine the social engineer's best-laid plans" (2002). In effect, human nature stymies tyranny. Accordingly, tampering with human nature is very risky: "Human nature shapes and constrains the possible kinds of political regimes, so a technology powerful enough to reshape what we are will have possibly malign consequences for liberal democracy and the nature of politics itself" (7).

He treats Aldous Huxley's *Brave New World* [(1932) 1969] as a cautionary tale. In that imagined world the state utilizes reproduction and gestation conditioning technologies to produce biological castes that provide the foundation for a very successful rigid social stratification system. Even if enhancement technologies are not dictated by the state but driven by consumer choice and the free market, Fukuyama worries that social inequality will increase and future rulers with superior enhancements will maintain insurmountable advantages over the ruled. Social and political mobility, so important to liberal democracy, will be restricted.

Conservationists reject Hughes' recommendation to use transtechnologies to help individuals become better citizens. Bio-social engineering, however well-intentioned initially, will eventually be used by the powerful to perfect domination. The critics of *civitas* transcendence take a position similar to Bill McKibben—don't risk this venture when what we have is good enough.

Transcendence nor Transgression?

In closing, I believe that the transhumanists and the conservationists do us a service by imagining the fate of subjects in a transtechnological world. Later in this book I'll introduce the debate over specific risks, but this frank exchange about the future of humanity is most satisfying in light of C. Wright Mills' questions. Nevertheless, we might consider the possibility that neither transcendence nor transgression will occur. Imagined futures need not diverge so much between the utopian and dystopian. A consistent skeptic would likely agree with Dan Quisenberry, the famous baseball pitcher, when he said, "The future is much like the present, only longer." New sociotechnical ensembles may emerge that are muddled and mixed. Perhaps it is not ascension or fall that we can expect, but continued struggle.

Chapter 3
Transformation of Body and Mind

Similar to exploring an old city I like to wander down the side streets of the transhumanity debate to read about fascinating areas of research and development such as artificial intelligence and artificial life. However, it's best not to get sidetracked. It helps to keep in mind that the debate is driven by basic concerns as much as it is by the science and technology. In particular, I am reminded of this by the personal revelations by authors. One scholar wonders whether he would remain consistent with his opposition to genetic engineering if his grandchild's life was in the balance. Another author relates his grief over the death of his parents. Others describe a chronic illness that they or loved ones must suffer and endure. These testimonials serve to remind me that the transhumanity debate is centered on the problem of human mortality.

The transhumanists are not the first to propose a way out of the problem, after all, most religions do this. It is the method they suggest that is controversial. They propose a solution in this life rather than in the hereafter. They pose a bold, rhetorical question: If the very constitution of the human body is what makes us and our loved ones susceptible to disease, decline, and death, why not transform it? In the past there was no reason to expect that such a thing could be done. Now there is.

A good deal of transhumanist writing describes how cutting-edge technologies may be used for transformation. I'm sure that this can be perceived as smart marketing—to persuade potential "customers" and investors of the feasibility of this venture. It is also a way for the particular writer to present his or her preference for a specific posthuman form or to criticize the model proposed by a peer. Some transhumanists favor a dramatic departure from human biology. I'll refer to this as radical transformation. Others recommend modest transformation which would entail retaining the basic human form albeit with augmentations.

Transhumanists are criticized for elevating consciousness over embodiment, but also for being too focused on the material basis of human existence. For anyone who has tried to get a handle on the debate, this can be quite confusing. Which is it? Both critiques make sense if we understand that the first is meant as an indictment of radical transformation and the second of modest transformation. I'll present the debate over radical transformation first.

S. Lilley, *Transhumanism and Society*, SpringerBriefs in Philosophy,
DOI: 10.1007/978-94-007-4981-8_3, © The Author(s) 2013

Radical Transformation

To possess ever greater powers while leaving the frail body behind is an audacious idea often attributed to mainstream religions or to cults. Faith in the intercession of a benevolent God is usually, but not always, a prerequisite. For instance, members of Heaven's Gate believed that the Earth was to be cleansed by extraterrestrials, whose presence in our solar system allegedly was hidden by the comet Hale-Bopp. By releasing their spirits into the protection of these aliens, they expected to enter the Next Level of existence. Members of The Raëlian Church, as well, believe in extraterrestrials as benefactors and the Church holds that a person's existence is extendable through cloning.

Transhumanists *disassociate* their movement from religions and cults. They make it clear that they do not appeal to supernatural forces (or aliens). In some ways this makes building a case for their audacious idea that much harder. Transhumanists *associate* their movement with science and engineering and therefore must abide by scientific-secular norms of persuasion. It is true that much of their work is speculative, imagining developments ten, twenty, or fifty years hence. Still, they provide plausible accounts of how to get there from here utilizing the language of science and engineering. Some of their predictions for technological innovations are near term and will soon be subject to verification. Overall, their methodology is similar to that practiced in the interdisciplinary field of future studies.

There is no way to prove that society and technology will proceed exactly along the lines predicted. Fortunately for the transhumanists, they do well simply by having others accept the possibility that the project is technically feasible. For example, the development of an efficient interface between brains and computers is a crucial step for radical transformation. There may be a dozen different approaches under construction or imagined, and although many will fail, it only takes one to produce the breakthrough. The transhumanists simply make the case that the odds are in their favor.

Mind over Body

We often refer to "mind" and "body" when describing the way we feel, for instance, my body aches after a grueling workout and my mind is tired from too much studying. The distinction between mind and body invites an interesting question, "What is the relationship between the two?" I've heard more than once from older relatives that their minds have remained relatively young but their bodies have aged. This places an emphasis on difference or independence between mind and body. When I hit an impasse in my writing, I hike on nearby nature trails to refresh my mind. This assumes a vital interconnection or interdependence.

Professionals who study this mind-body duality similarly can pursue the independence or interdependence interpretations. In January 2008, Miguel A. L. Nicolelis, a neuroscientist at Duke University, provided a startling demonstration of independence that suggests that the mind can bypass the body. Every scientific experiment has production pieces and these are the key ones in this demonstration: a monkey (named Idoya, in North Carolina), behavioral training, implanted electrodes, a treadmill, a 200-pound, 5-foot humanoid robot (named CB, in Japan), a signal relay system using computer and transmission technologies, and a movie screen. This is how Sandra Blakeslee (2008), the New York Times science correspondent, described the event:

> As Idoya's brain signals streamed into CB's actuators, her job was to make the robot walk steadily via her own brain activity. She could see the back of CB's legs on an enormous movie screen in front of her treadmill and received treats if she could make the robot's joints move in synchrony with her own leg movements. As Idoya walked, CB walked at exactly the same pace. Recordings from Idoya's brain revealed that her neurons fired each time she took a step and each time the robot took a step…
>
> An hour into the experiment, the researchers pulled a trick on Idoya. They stopped her treadmill. Everyone held their breath. What would Idoya do? "Her eyes remained focused like crazy on CB's legs," Dr. Nicolelis said. She got treats galore. The robot kept walking. And the researchers were jubilant.

Idoya's accomplishment is taken as proof of concept that the mind (or brain), suitably connected, can direct devices. The demonstration also succeeds by persuading the audience to consider Idoya's body (walking on the treadmill) and CB's body (also walking on a treadmill) as interchangeable objects under her mind's control. Research in brain machine interface has relevance for individuals who have suffered amputation or spinal injury, with work underway on how to allow human subjects to operate an artificial limb or exoskeleton by thought. Here again the mind is understood to be the active and independent agent.

Individuals who wish for restoration of mobility can find hope in Miguel Nicolelis' statement that "The body does not have a monopoly for enacting the desires of the brain." (Blakeslee 2008) Many transhumanists also take heart in new possibilities emerging from the confluence of neuroscience, computing, and robotics. They readily conceptualize the human body as one substrate for the mind, dispensable once better replacements are engineered. Prosthetic limbs, artificial hips and knees, cochlear implants, and pace makers are standard medical devices. Artificial bone, tendon, muscle, skin, blood, etc., are commercially available or are being developed. Each component may be seen as just one exception to an otherwise natural form, however taken together one is left with the impression that the organic body is replaceable. At the very least, it makes you wonder, "What is so special about human biology?"

The research described above imagined along the lines of science fiction characters such as the bionic man, bionic woman, and Robocop suggest that human biology is not so special and that a synthetic body would do just fine. Consider for the moment the alleged advantages: It would be more durable and not prone to disease and senescence and it could be upgraded as the technology improves. It would allow for more power, agility, and functionality. If the organic brain is seen as a limiting factor, "uploading" or copying the mind to a more promising medium would be the next step. That level of technological prowess, if ever reached, opens up even more radical options such as digitizing the mind to allow existence in virtual worlds, thereby leaving behind human biology altogether.

Of Substrates and Cyborgs

What understanding of you and I—of human subjectivity—allows for the assumption that the human body is a nonessential element to one's being? Let's start with Andy Clark's portrayal of humans as "natural-born cyborgs." The term "cyborg," an abbreviation for cybernetic organism, denotes an entity part biological and part machine. Manfred Clynes and Nathan Kline (1960) first used the term to describe an astronaut capable of surviving a lengthy period of time in space: built-in monitors and drug infusers, artificial lungs, etc. Clark, a cognitive scientist, asserts that humans have always been cyborgs, in the sense of incorporating our best creations as a way to extend our reach. He identifies "cognitive hybridization" as the quintessential feature of our humanity:

> [I]t is our special character, as human beings, to be forever driven to create, co-opt, annex, and exploit nonbiological props and scaffoldings. We have been designed, by Mother Nature, to exploit deep neural plasticity in order to become one with our best and most reliable tools. Minds like ours were made for mergers. Tools R-Us, and always have been (2003:7).

Some of the props and scaffolds that the mind utilizes for expansion include older technologies, for instance, paper and pen, and newer technologies, for example, computers. Think of Google and the extension it provides for our inquisitive mind or the Nature or Discovery channels on television.

A trek through the Amazon basin would be arduous to say the least, but by watching a documentary on the rainforest you can get a sense of the environment without leaving your home. Watching a documentary is not the same as being there, although the difference will diminish with new simulation mediums. Consider what Miguel Nicolelis and his team had Idayo do. My children do something similar when they play Nintendo's Wii game system. They move a character through an environment on the screen and through audio, visual, and force feedback become, to a certain extent, the character within the environment. Technologists are working on improving haptics and telepresence. If reality simulation, rather than game play, is the goal this is how it might work: The explorer at home wearing specialized gear remotely controls an agile robot through the rainforest in a fashion similar to NASA engineers piloting the Mars Rover. Even better than the Rover, the robot returns a rich stream of sensory information (audio, visual, as well as tactile, olfactory, etc.) to be experienced by the tele-traveler. The not-too-intrepid explorer feels immersed in the environment.

In terms of exploring harsh terrains, the body may be an encumbrance or a liability. It certainly is impractical for exploration of the deep sea, volcanoes, and other planets. According to Clark, we find the "good old-fashioned skin-bag," as he refers to the body, too restrictive for even ordinary purposes such as talking to family and friends over distance. We utilize communication mediums all the time to project our voice. Andy Clark treats the body like any other prop or scaffold. He implies that the genome-governed components that continue to serve the mind in important capacities in the environments that we explore and construct will be retained while others will be replaced or augmented by new props. The body has no special dispensation from techno-cultural selection pressures.

Kevin Warwick, a self-described cybernetic pioneer, has used his own body as a site for cyborg experimentation. With implants connected to nervous fibers in his left arm and a radio transmitter/receiver sending signals from his nervous system to a computer he operated a robotic arm and exchanged electrical signals with his wife (similarly equipped). Brain implants would be so much more powerful. (They have been tested on individuals suffering severe hearing loss or paralysis but the surgical procedure is risky.) Warwick welcomes the therapeutic applications but he also imagines posthuman capabilities:

> At present our method of communication, speech, is very slow, serial and error prone. The potential to communicate by means of thought signals alone is a very exciting one. We will probably have to learn how to communicate well in this way though, in particular how to send ideas to one another. It is not clear if I think about an ice cream are my thoughts roughly the same as yours - we will have to learn about each other's thoughts. Maybe it will be easier than we think, maybe not. Certainly speech is an old fashioned, out dated means of communication - it's on its way out! (2008)

Whether in terms of Warwick's bypass system or Clark's new prostheses, the body diminishes in importance. The mind, in contrast, gains more degrees of freedom. According to Clark this will accelerate:

> [N]ew thinking systems create new waves of designer environments, in which yet further kinds of extended thinking systems emerge. By this magic, seeded long ago by the emergence of language itself, the ratchets engage and the golden machinery of mind-design, mind redesign, and mind re-redesign, rumbles into life. The process continues, and it is picking up speed. Some of our best new tools adapt to individual brains during use, thus speeding up the process of mutual accommodation beyond measure. Human thought is biologically and technologically poised to explore cognitive spaces that would remain forever beyond the reach of non-cyborg animals (197).

Simon Young (2006) pronounces Homo cyberneticus to be the next stage in human evolution. He traces cyberneticus to the Greek, *kubernetes*, or steersman of a ship. He understands the mind to be the steersman and the body to be an unworthy vessel. He asserts, "The body may want to self-destruct—but does the mind? No. Yet our genes insist upon it, against our will." (371) Freed from "genetic slavery," minds will evolve, thereby setting the stage for these cognitivist triumphs: 1) the evolution of a cybermind emerging from the network of inter-dependent minds (318), 2) "the mind of evolution become conscious of itself" (39), and coming to know the "Mind of God."(367)

Neurobiologists assert that the mind is an epiphenomenon of the working brain. It is dependent on the functioning of cells, genes, hormones, proteins, and blood.

This presents a challenge to the conception of mind independence from body. Ray Kurzweil offers a "patternist" perspective as an alternative to this materialist account. Biological and nonbiological systems such as computation systems need not be treated as being fundamentally different because all systems are reducible to patterns. Evolution, biological and technological, entails the progressive development of patterns. Human intelligence is an evolutionary milestone that Kurzweil readily admits has a biological basis, nevertheless as its pattern is made known through neuroscience, informatics, etc., ways will be found to replicate it or reformat it to allow symbiosis or mergers with newly created forms of computation and artificial intelligence. The pattern that is intelligence will continue to evolve. Like Young, he foresees nothing, including the organic brain, getting in the way: "[I]ntelligence is the most powerful "force" in the universe. Intelligence, if sufficiently advanced, is, well, smart enough to anticipate and overcome any obstacles that stand in its path." (2005:206).

Religious Critique: Escape the Body, Lose the Soul

Instead of attacking transhumanism with a scientific-skeptical question, Can this be done?, opponents challenge with a moral question, Should this be done? Most non-experts willing to hear out both sides will most likely decide their positions based on the debate over the latter question. The preference for mind over body has its religious parallel in the elevation of spirituality over material existence. Platonism, Manichaeism, and Gnosticism, for example, hold that the material world is a corruption of an ideal state. Humanity, trapped and confused in the physical realm, nonetheless yearns for goodness and to rejoin the divine being. Redemption is possible by a renouncement of physical attachments and through special knowledge or aesthetic practices that promote spirituality.

Irenaeus, and other central figures behind the institutionalization of Christianity and the establishment of Church doctrine, strenuously opposed Gnosticism and campaigned for the exclusion of the Gnostic Gospels in the canon. He argued that a non-divisible God oversaw creation and intended human existence to be the way it is. Humans develop spiritually through living and suffering, not by trying to escape life. Similarly the incarnation of Jesus is thought to be a pivotal act in salvation.

Erik Davis (1998) coined the term "techgnosis" for what he characterizes as a contemporary, secular version of Gnosticism. Pure intelligence or information replaces spirituality as the ideal state, technology replaces the role of God or Christ as savior, but otherwise the message is the same—the body is an impediment to humanity's quest. C. Christopher Hook accuses transhumanism, specifically, for promoting Gnostic claims:

> Transhumanism is in some ways a new incarnation of gnosticism. It sees the body as simply the first prosthesis we all learn to manipulate. As Christians, we have long rejected the gnostic claims that the human body is evil. Embodiment is fundamental to our identity, designed by God, and sanctified by the Incarnation and bodily resurrection of our Lord.

Unlike gnostics, transhumanists reject the notion of the soul and substitute for it the idea of an information pattern (2004).

Elaine Graham contrasts transhumanist anthropology with "theological anthropology":

This predilection for the qualities of detachment, omniscience, immutability and incorporeality translates into a transhumanist anthropology founded on disdain for embodiment, vulnerability and finitude in which only the fittest will survive. A theological anthropology, on the other hand, would see things differently, however, not least in its eschewal of a symbolic of transcendence premised on omnipotence, immortality and rejection of the material world. This vision regards the immanent, material world not as an impediment to authentic spirituality, but the very realm of divine-human encounter (2003:40).

Henk Geertsema asserts that "We are called to respond to God, who created us, and to live according to the intentions given with creation." (2006:313) God fashioned man and woman from material to take a specific physical form. This is not a mistake or punishment, rather it is a gift. God is pleased with His creation and, despite pain and toil, existence in the human form entails a state of grace.

Christian critics contend that becoming a full-fledged cyborg is perilous. First, the modifications and enhancements will be done in defiance of His will. Second, the excursion into human–machine genesis will be done without the wisdom of God. Will the right balance be struck? It is possible that in the attempt to isolate and merge human intelligence with artificial intelligence, for example, vitality, spontaneity, and emotionality will be reduced. Third, and most importantly, radical transformation means radical alienation from God. It is akin to a second Fall, a repudiation of God's gift of creation and a prideful decision to go it alone. Mark Hanson describes this in terms of sin:

Only a faith that recognizes a dependence upon God can save the self from the sin that is the human attempt to make the self God...Sin is occasioned by anxiety when a person fails to acknowledge his or her finiteness and the dependence of his position upon God and thereby seeks powers and securities that transcend the possibilities of human existence (1999).

Christians find in the New Testament hope for a different kind of transformation through the resurrection of Christ. The gospels describe Jesus rising from the dead, having physical form, and relating in ordinary ways with others, e.g., sharing meals with his disciples. A promise is made that those who believe in Christ and follow his way will also have eternal life. Robert Song (2006) states that for Christians, "transcendence takes the form not of escape from the vulnerability that accompanies embodiment, but of the hope of a transformed body in a divinely renewed heaven and earth." This transformation does not entail technological intervention, nor is it accomplished through the mind's liberation from the body. Rather, it is bestowed by the redemption of Christ and it involves a perfection of the original form.

What if the transhumanists are wrong to devalue the human body? What happens to us if we work in defiance of God's plan? What if by radically altering our existence we lose a state of grace or the chance to exist in the afterlife in perfect form and in close relationship to God? These questions are meant to warn others, especially the faithful, that radical transformation is a dangerous path to take and that there may be supreme costs associated with it.

Secular Critique: Escape the Body, Lose the Self

Secular critics present a cautionary tale of radical transformation that is closer to home, literally. It has to do with kids (and adults) playing on the computer. In many countries, the average numbers of hours spent online has increased over the past ten years. As internet use increases there appears to be a reduction in television viewing time, a tradeoff that many would find acceptable. Some studies have found, however, that time displacement is more serious in that family and community involvement declines. (See, for example, Nie and Hillygust 2002)

Social scientists and medical professionals are most concerned with ill effects associated with heavy internet use. Heavy users often neglect studies, work, family, and friends. They often lose sleep and do not eat well or exercise. (Block 2008) Although time spent can accumulate from a variety of online activities, for example, chat, social networking (Myspace and Facebook), gambling and pornography, some researchers have identified online gaming as a particular problem. A recent study of fourth, fifth, and sixth grade Korean school children found that gaming, but not other online activities, was associated with a perceived decline in family time and family communication. (Lee and Chae 2007) (South Korea is a leader in broadband access and high school students in that nation average 23 hours per week on gaming. (Kim 2007)).

What makes gaming so captivating? According to Anna Meenan, massively multiplayer online role-playing games (MMORPG), such as World of Warcraft, are structured to require increasing playing time to achieve objectives: "At the highest levels, players must band together into guilds to go on quests or raids that can require ten or more hours of continuous play, with some players reporting playing over 70 h per week." (2007:1117) Besides the game commitments, some researchers cite the allure of virtuality. In her groundbreaking investigations into Internet play and identity, Sherry Turkle (1995) found that some of her interviewees simply preferred their virtual self over their real self and wanted to spend as much time as they could in their virtual world, and this was when such worlds and selves were manifested through text! Today, the graphic worlds are so much more sensory satisfying and avatars provide for a better presentation of self.

Barbara Becker (2000) cites one person admitting in her study that gaming is a way "to escape from the bodily prison." She believes that this is a common motivation. She notes that the material world and physical body entail stubborn realities and that "[i]n communicating with the world through technology or media, people try to avoid feeling the concrete resistance of its materiality." Switching to virtual reality provides a sense of relative gain in mastery: "as a result the power of the subject seems to increase. The subject solipsistically establishes itself as the ruler of a world." Becker asserts that the transhumanist vision of radical transformation is similarly seductive, promising many more degrees of freedom to play and self-experiment. She dismisses this for being "a reconstruction of old fantasies which are returning in new technological clothes and making a great deal of noise."

One such "old fantasy," according to John Sullins (2000), comes from Descartes' philosophy. Rene Descartes treated the mind and body as being distinct. The body, but not the mind, is of the physical world, influenced by natural laws and operating in a similar fashion as machines. Through the body's sensory receptors, information is presented for the mind's perusal. In his famous thought experiment Descartes imagines a demon manipulating the senses. He also offers ordinary examples of sensory error. The lesson to be drawn is that the body is not to be trusted as a source for certainty regarding one's existence. Only the action of the mind, or the "I" that is thinking, is certain. The primacy of the mind is clear in his epistemology that favors rational thought and deduction.

Sullins applies Maurice Merleau-Ponty's phenomenology to point out, how-ever, that in our lived experience we do not treat the mind as the independent subject and the body as its object. As we go about our daily lives we do not instruct every bodily activity, nor do we perceive our bodies as being distant from us. Furthermore, body image, physical habits and sensitivities, etc., are crucial to our experience of self. The body is not an object, rather it is intricately involved with the mind and constitutive of the subject. Only upon reflection or in the develop-ment of philosophical systems do we disassociate the body and mind. Sullins asserts that an acknowledgment of "body-subject" has implications for radical transformation:

> Our personal identity is based on the fact that we are embodied in a particular way and have experienced a certain, reasonably continuous, spatio-temporal history without which we have no identity, we are not a person. Thus the technologies we have been discussing will not be able to deliver on their more ambitious claims. We will not be able to upload our mind into a machine and still remain ourselves for long. Even if uploading our consciousness into a machine was somehow technologically feasible, all we would achieve is the slow annihilation of our personality as it melted into the functions of the machine over time. At best we would create a new machine personality with a new distinct individuality.

What if an organic body or machine substrate could be avoided altogether? What if our pattern could be liberated? N. Katherine Hayles (1999) notes that such speculation is encouraged by a strategy employed in cybernetics to treat mind as an information system irrespective of platform or housing. Neuron or computer chip, flesh or metal, it does not matter. She argues against the reductionism and asserts that "for information to exist, it must *always* be instantiated in a medium." (13) Like Sullins, she insists that whatever mind might be generated on another medium it will not be human: "Human mind without human body is not human mind." (246).

Transhumanists, in particular, have an interest in promoting the idea that cyborgs develop more capacities over time and in the future they will span more domains and will have more complex interconnections. Invariably the imagery used to describe this being is a distributed network. What will happen to the human self? Will a central agent exist or will it be distributed or dissolved in the network? These are very difficult questions primarily because there is no consensus regarding the nature of the human self. Andy Clark favors Daniel Dennett's theory that the central self is an illusion.

> There is no self, if by self we mean some central cognitive essence that makes me who and what I am. In its place there is just the "soft self": a rough-and-tumble, control sharing coalition of processes– some neural, some bodily, some technological– and an ongoing drive to tell a story, to paint a picture in which "I" am the central player (138).

If this is the case, there is no reason to be overly concerned about the loss of a sense of self in transhuman existence. Configurations that provide identity will emerge from a more extensive network of advanced processes just as surely as they have emerged from a more limited "ensemble of tools." Kurzweil believes that we are sufficiently pliable to retain continuity of identity as we change. Hayles, however, does not preclude negative outcomes. She takes seriously a concern raised by Norbert Wiener, a principle architect of cybernetics, that the subject may be subsumed. Estrangement is very possible. Exploitation and manipulation need to be considered as well: "The ultimate horror for the individual is to remain trapped "inside" a world constructed by another being for the other's own profit." (162).

Transhumanists urge men and women to take charge of their own destiny, to reject biological limitations, and to take on new capacities. They give every impression that intrepid innovators will remain in control of this process. However, Hayles, Sullins, and others question whether the willful agent will be left intact as the body is left behind. Will the self survive radical transformation?

Moderate Transformation

Gregory Stock, a biophysicist, believes that the public will not buy radical transformation. This can be understood literally: Consumer demand drives the commercialization of technology and consumers will opt for safe, reliable, and reversible procedures over those that are exotic, risky, and permanent. Why put your mind, body, and self in jeopardy if relatively benign technologies can provide good health and longevity? Why become a cyborg when being a "fyborg" offers the same benefits without the costs? Alexander Chislenko (1995) defined a functional cyborg or fyborg as a biological organism supplemented with technological extensions. Many of us utilize extensions such as eyeglasses, contact lenses, hearing aids, blue tooth head sets, IPods, and cell phones, so according to the definition we are fyborgs. With further miniaturization of electronic/computer devices more and more gadgets will become wearable.

Moderate Transformation as Value Gained

According to Stock, "people want to be healthier, smarter, stronger, faster, more attractive" (2002:179) but most people are technology pragmatists rather than technophiles. They also tend toward conservatism when it comes to the human body. A chapter entitled, "Our commitment to our flesh," best expresses his contention that if given a choice between hardware implanted in tissue and equally effective wearable devices, most people will choose the latter. A powerful storage and retrievable system that can be worn poses fewer health risks, and is easier to repair, upgrade, and replace than memory chips embedded in the brain (25–26). Moreover, a wearable device does not threaten the sense of body integrity. He sees no value in Kevin Warwick's cyborg project when there are more practical alternatives.

Fyborgization cannot satisfy all desires, especially for better health and longevity, so Stock recommends biomedical augmentation as well. He favors gene therapy and germline selection and modification, in particular. The accepted wisdom today is that the expression of genes is important to health and longevity. The goal of gene therapy, in simple terms, is to replace dysfunctional genes with functional ones. As it is practiced now, somatic cells are targeted, e.g., liver and lung cells. Changes in these cells cannot be passed along to progeny. Germline gene therapy, on the other hand, involves germ cells (e.g., egg, sperm) or cells of the blastomere. Ever the pragmatist, Stock prefers germline intervention because of certain advantages. Most important is the timing of the intervention. Germline engineering is done at the very start of biological life. At this stage, germ cells or embryonic stem cells are amenable to laboratory techniques. Also, modifications made at this stage may be carried on to every cell in the body. Intervening later (with somatic cell gene therapy) entails dealing with a much more complex organism and under a clinical setting. Moreover, bioengineers must create a safe and effective vector system, typically a genetically-modified virus, to target specialized cells.

The downside of germline engineering's thoroughness is that deleterious modifications would be systemic and possibly passed down to progeny through sexual reproduction. Critics point out that we are far from understanding the complex actions and interactions associated with genes and that experimentation along these lines would be imprudent and unethical. Stock suggests a technological fix: synthetic chromosomes with reversible genes. Gene modules on an extra chromosome (loaded at the point of germline intervention) could be switched on by taking a tailor-made drug and, if ill effects were detected, switched off. This is how Stock sees it playing out across generations:

> Imagine that a future father gives his baby daughter chromosome 47, version 2.0, a top-of-the-line model with a dozen therapeutic gene modules. By the time she grows up and has a child of her own, she finds 2.0 downright primitive…The daughter may be too sensible to opt for some of the more experimental modules for her son, but she cannot imagine giving him her antique chromosome…As far as reverting to the pre-therapy, natural state of 23 chromosome pairs, well, only Luddites would do that to their kids (76).

From Stock's point of view, the beauty of this approach is that the novel gene modules could boost immunity, thwart cancer, and slow down aging without necessitating a radical departure from the human form. Moreover, this prevents disease whereas other methods including somatic cell gene therapy and therapeutic stem cell techniques are geared toward treating disease. He acknowledges that augmentation of the genome and "entry of laboratory machinery into human reproduction" (111) will take some getting used to, nonetheless, given mainstream acceptance of other assisted reproductive technologies and the clear advantages of germline engineering, he is optimistic of its success.

Moderate Transformation as Value Lost

Stock's transhuman may be recognizable to us, nevertheless, critics object to its bioengineering especially when it takes place at the start of life. For religious conservatives any departure from what they believe to be the God-given constitution of human nature goes against His will. Those who oppose assisted reproductive technologies and abortion also oppose germ line engineering. Conception and gestation are deemed to be inviolable, a sacrosanct beginning that is meant to proceed naturally.

Arguments aimed against radical transformation are used as well to contest moderate transformation. Scientists may try to "play God" but they do not possess His wisdom. There is too much preoccupation with temporal matters and not enough attention paid to spiritual development. What good is the augmentation of the body if it comes at the expense of the soul?

Stock and others present moderate transformation as the means to bountiful longevity, but so-called "life cycle traditionalists," such as Leon Kass, assert that there will be costs. He believes that the normal human lifespan promotes a more focused approach to life projects. He warns of ennui at the personal level and generational conflict over finite resources at the societal level.

Francis Fukuyama sees sexual reproduction as a genetic lottery that acts as a social equalizer. With controlled reproduction in a market economy, biological advantages will be purchased. Paul Lauritzen wonders how a bio-privileged class will relate to the unenhanced or poorly enhanced. Given that chronic illness and senescence takes a toll on physical appearance and productivity, will they feel pity or disgust for those who can't afford augmentation or elect to forego it? He worries that the significant differences in life experience and longevity "may erode a sense of common humanity" and "run the risk of blocking compassion and advancing intolerance." (2005: 30).

Secular critics reject Stock's libertarian argument that germinal choice technologies will allow for greater freedom of choice. Instead, they see parents gaining ever more control over their "designer babies," social competition pushing people to alter their bodies, and biomedical corporations manipulating consumers through advertising and marketing campaigns.

I have presented only a sampling of the objections to moderate transformation, nevertheless this selection should be sufficient to underscore the basic point made by critics that the application of transtechnologies can never be moderate in its effects. Personality systems, micro and macro social systems, according to these critics, have evolved according to certain natural parameters and change outside these parameters will necessarily cause disturbances. Many harbor a deep suspicion of what is sometimes referred to as "corporate science" and do not want to see further market penetration into human life. Rather than focusing on what might be gained through transtechnologies, they highlight personal and collective goods that might be lost.

Leon Kass states it this way: "We need to realize that there is more at stake in the biological revolution than just saving life or avoiding death and suffering. We must also strive to protect and preserve human dignity and the ideas and practices that keep us human." (2002:1) This emphasis on human dignity or human flourishing is apparent in Michael Sandel's call for an "ethic of giftedness" (2007:45) Sandel celebrates the openness of human life, unenhanced. As long as an athlete is not artificially enhanced or a child is not designed, we can appreciate the unique qualities and achievements that he or she brings to the world. Conversely, striving for perfection by means of bio-engineering entails an "excess of mastery and dominion that misses the sense of life as gift." (62) Humility, empathy, patience, and wisdom are fostered through acceptance of limitations.

Bill McKibben (2003) also believes that enhancements, paradoxically, will stymie human flourishing. He warns that the designed person will be crippled by existential uncertainty, never quite sure if the "programming" is responsible for his or her motivations, always wondering "why I choose what I choose." (49) Achieving "flow"–a remarkable experience of being completely immersed in a challenging activity—is unlikely given such self-doubts.

Leon Kass agrees that nobility or elevation of character will not be fostered by transgenic technologies. Life may be extended, suffering reduced, but with the grit of human life removed—the hardship, the mortality—there will be little opportunity and inclination to face and overcome adversity and thereby achieve true dignity. He declares that "the downward pull of bodily necessity and fate makes possible the dignified journey of a truly human life." (2002:18) The natural parameters of the human species, including the genome, sexual reproduction, the life cycle, generations, and kinship, not only provide the best context for flourishing, they serve as a common heritage and mooring. McKibben (2003) predicts that the first enhanced child will "see a gap between himself and human history" (64) and "[h]e'll be marooned forever on his own small island, as will all who follow him." (65).

Defending Posthuman Dignity

Transhumanists reject any argument derived from a theological claim of telos, i.e., that human beings were fashioned in a particular form to realize a divine purpose. Such a claim presupposes that 1) a benevolent supernatural being exists, 2) that this being has specific expectations for humanity, and 3) that these have been made clear to humanity. Ordinarily a believer makes a leap of faith that such a revelation has been made as truly related by a holy figure or recorded in a sacred text. The transhumanists practice secular-scientific skepticism and dismiss creation stories, such as the one told in Genesis, as lacking scientific validity. Furthermore, they insist that any proposal regarding science and technology policies based on creationism or Intelligent Design should not be taken seriously in the public arena.

The transhumanists contend that the "bioconservatives" have it all wrong about human nature because they overemphasize stasis. Andy Clark explains that not only has the species changed over time through biological evolution, it is unusually dynamic in other ways:

> It is our natural proclivity for tool-based extension, and profound and repeated self-transformation, that explains how we humans can be so very special while at the same time being not so very different, biologically speaking, from the other animals with whom we share both the planet and most of our genes. What makes us distinctively human is our capacity to continually restructure and rebuild our own mental circuitry, courtesy of an empowering web of culture, education, technology, and artifacts (2003:10).

In his article, *"In Defense of Posthuman Dignity,"* Nick Bostrom (2005) questions whether human dignity is fostered through conservation by challenging an underlying premise that human nature is set or pinned down by the human genome. He insists, instead, that ours is a species that extends and transcends biology through social and technological constructions, and as these change, we change, generation after generation. There is no stable state to preserve:

> What we are is not a function solely of our DNA but also of our technological and social context. Human nature in this broader sense is dynamic, partially human-made, and improvable. Our current extended phenotypes (and the lives that we lead) are markedly different from those of our hunter-gatherer ancestors…
> Yet these radical extensions of human capabilities– some of them biological, others external – have not divested us of moral status or dehumanized us in the sense of making us generally unworthy and base. Similarly, should we or our descendants one day succeed in becoming what relative to current standards we may refer to as posthuman, this need not entail a loss dignity either (213).

Simon Young (2006) identifies in human beings the "will to evolve" toward "ever-increasing survivability and well-being." (19) Fortunately, according to Young, humans are uniquely gifted to exercise this will. We have bypassed Darwinian evolution with the first steps in "designer evolution" and continued progression is our species' destiny and the individual's best chance for flourishing. Attempts to conserve human nature by banning enhancement will violate the quintessential drive of human beings! Ray Kurzweil suggests a cosmic destiny: "As I see it the purpose of the universe reflects the same purpose of our lives: to move forward to greater intelligence and knowledge. Our human intelligence and technology form the cutting edge of this expanding intelligence…" (2005:420).

Taboo or Tolerance

Conservationists and transhumanists argue over what we should become on the basis of their respective understanding of what we are. What is the essence of human nature? If we come to the debate already convinced of the answer it will dictate our response to transhumanity. For instance, if one believes that there is a divine purpose behind the creation and preservation of human biology, a transhuman turn would be understood as contrary to God's will. Such hubris to transform human beings! If one believes that humans are creatures of evolution with the capacity and drive to evolve in new ways, a transhuman turn would be understood as a new step along the same path. Carry on!

However, not everyone is so certain about the essence of human beings. What if we are suspicious of the very notion that there is one true nature or we are simply uncomfortable with the debate at such an abstract level. For the practical-minded, their stand vis-à-vis transformation may come down to a cost-benefit analysis. As I'll describe in the next chapter, the transhumanity debate offers a well-developed discussion of risks. However, is there a more intuitive way of knowing what to do?

A while ago I was talking to a neighbor about my studies on transhumanity and when we reached the subject of genetic engineering she shook her head and said, "That really bothers me if they mess with genes… That's very creepy." I suspect that this is a common sentiment. Leon Kass argues that we should attend to the "Yuck Factor" because the feeling of repugnance registers the violation of a taboo and serves as a warning of overstepping our bounds. He states that "repugnance is the emotional expression of deep wisdom, beyond reason's power completely to articulate it" and that "revulsion may be the only voice left that speaks up to defend the central core of our humanity." (2002: 147–150) He contends that most people are deeply troubled by the prospects of human bioengineering and that laws and regulations should reflect that.

Transhumanists strongly disagree, and counter with a rights-based argument. They note that especially in heterogeneous societies, citizens are unlikely to share the same perspectives and beliefs. What one person perceives as strange and dangerous another might see as potentially uplifting. Nick Bostrom (the Founding Chair of the WTA) insists that transhumanists do not wish to impose new technologies on anyone, rather they are simply requesting ample latitude for individuals to choose scientific and medical advances:

> "[t]ranshumanists promote the view that human enhancement technologies should be made widely available, and that individuals should have broad discretion over which of these technologies to apply to themselves (morphological freedom), and that parents should normally get to decide which reproductive technologies to use when having children (reproductive freedom)" (2005:203).

Citizens may find other citizens' religious beliefs and practices erroneous, strange and, perhaps, offensive, but tolerance is expected for the sake of the entire system or, if for no other reason than to ensure quid pro quo. If social conservatives wish to make basic life decisions without interference, so too should they allow others this autonomy. Gut reactions are insufficient grounds to deny another person the right of self-determination.

Of course, transhumanists understand that if enhancement technologies are perceived by most to be out of bounds, prohibition is more likely. Instead, they portray these technologies as being comparable to protected procedures such as cosmetic surgery, laser eye surgery, sex reassignment therapy, psychotropic medication, hormone treatments, and physical fitness training—all means by which some individuals pursue the right to modify their bodies and minds. Individuals exercise reproductive or procreative rights through mate selection, sperm and egg selection, and embryo screening, and Gregory Stock (2002) describes germline engineering along the same lines. Transhumanists want transgenic applications to be treated no different than extant treatments: face safety and efficacy assessments, but not a "yuck trial" or religious litmus test.

Chapter 4
Rhetoric of Risk

Some individuals can't bear the thought of a transhuman future. They can't see a place for themselves among the transcended and transformed. Nor can they see transhumanity supporting their beliefs and values. They feel an anticipatory anomie—an expectation of being displaced from a changing culture and society. The possibility of physical harm, the type of risk that comes quickest to mind, is not a primary consideration. In other words, absent personal risk they still would oppose transhumanity. Others can't bear the thought of a future without transhumanity. Their hope for self-transformation is kept alive by this prospect. It would be better if risks were low, but they would accept elevated risks in light of expected rich returns.

How accurate could a risk analysis of transhumanity be, anyways? Common sense informs us that confidence in predictions diminishes as the timeline extends out and the number of variables increases. Both sides figure transhumanity would require global sociotechnical change over a period of decades! How can anyone foretell with any precision the risks involved in that?

If subordinate to convictions and not a viable actuarial exercise, why are both sides so intent on characterizing future risk? I believe it is done for the sake of rhetoric. Please don't take this as sarcasm. By "rhetoric" I'm not using the pejorative connotation of word play which distracts and misleads. Rather, I mean argumentation of a very serious nature meant to influence the thought and conduct of an audience or audiences. Whether small (a government committee) or large (the literate adult population), it is important to win over audiences because therein lies important resources such as votes and financial support.

Although individuals live in the present, they have a stake in the future and act to the best of their ability to secure a promising future. Much of this is mundane. We pay utility bills for monthly service and schedule periodic health check-ups. All this is done with the assumption that present regular patterns will continue into the future. Nevertheless, we also anticipate that change will occur, personal and societal. For example, venture capitalists invest in technologies that they hope will have commercial success ten years down the road. Expecting a devastating nuclear

S. Lilley, *Transhumanism and Society*, SpringerBriefs in Philosophy,
DOI: 10.1007/978-94-007-4981-8_4, © The Author(s) 2013

war, survivalists build personal fallout shelters. Although these futures are imagined or have yet to be realized (let's hope nuclear warfare never is), the commitments are very real. Certainly the businesses involved benefit.

Imagined futures are socially constructed and shared. Purveyors work through social networks and the mass media and compete with one another. Louise Bedsworth, Micah Lowenthal, and William Kastenberg (2004) note that in the California low-level radioactive waste debate opposing parties attempted to influence public opinion by portraying different futures, one safe and the other hazardous, each reinforced by its own "rhetoric of risk." The transhumanity controversy is more extensive, nevertheless, it is conducted in a similar manner. If the transhumanists dazzle audiences with alleged future benefits and go unchallenged, they will secure commitments for transhumanity. Similarly, if opponents face no opposition as they depict a dark and dangerous transhuman future, they will garner sufficient resistance. Obviously, neither party can allow the other to unilaterally characterize risk. It does not matter if predictions of boons or catastrophes are necessarily deficient of evidence because preemption, not proof, is the goal.

The Social Construction of Risk

For the professionals employed to minimize hazards associated with technologies it is best to treat risk an as objective phenomenon—discrete, measurable, and predictable. Designers, for example, treat risk as an inherent quality of a material or operational state, for example, the stress capacity of steel. In order to minimize risk, bridges are built with steel girders of a specified grade, automobiles are equipped with air bags, and nuclear power plants are regulated by fail safe operations. However, we know from bridge collapses, automobile fatalities, and nuclear accidents that risk is hardly eliminated. Science and engineering are very much about control, nevertheless uncertainty, complex/chaotic dynamics, disagreement, and competing social interests are often present at the design stage and at the use level. For instance, in capitalist economies technologies need to be affordable (to entice customers and compete with rival producers) and compromises often are made in order to keep costs down. Governments allow and sometimes promote dangerous technologies in the name of national security or economic competition. Risk is governed by technical *and* social factors.

STS scholars emphasize that risk is socially defined and that such work is accomplished through language, power, organization, and culture. Often it is reconstructed *following* a devastating system failure. After the space shuttle disasters, for instance, investigators pointed out (and sometimes disagreed) on the alleged operational and material flaws. The Deepwater Horizon oil spill of 2010 brought public attention to blowout preventers and chemical dispersants with experts disagreeing on health risks of the latter. Besides such *ex post facto* inquiries we are also finding an increase in proactive risk characterization, as exemplified by the low-level radioactive waste debate. Inquiries and debates increasingly take place in the public domain.

Brian Wynne and Sheila Jasanoff were among the first researchers to investigate social controversies over risk. Jasanoff (1995) focused on the role that experts play and Wynne (1987) studied laypersons' involvement. They found vigorous contests in which contestants often disagree about the relevant evidence or methodology and question each other's claim to authority. This appears to be especially true in America given the culture of litigation and the often adversarial relationship between industry and advocacy groups. (Parthasarathy 2004) There may have been a time when deference to scientists and engineers on matters of technology and risk was assured, but not today. By means of advertising campaigns and court proceedings, advocacy groups publicly challenge expert risk assessment on technologies as diverse as genetically modified organisms, high-power transmission lines, and vaccines.

Risk and Social Movements

Cynthia Selin explains that "the future is a rhetorical and symbolic space" used to work out what is a new technology "but it also serves a productive role that underlies everyday decision making, alliance building, and resource allocation." (2007:215) Movement leaders take care to characterize risk in a way consistent with the world view and value set held by their members. Social movement organizations (SMOs) depend on members for financial support and participation in events, and leaders do well to please and not alienate their base (see Zald and McCarthy 1979). After all, social movements succeed depending on the assets brought to bear. Another way to look at this is in terms of the integrity and consistency of the organization's *raison d'être* and character over time. Through the trials of initial formation and various policy campaigns, organizations develop a core identity or history. Leaders must be careful not to deviate from that when formulating the organization's public presentation of risk.

Assessments of the risks associated with transhumanity differ dramatically because they reflect the advocacy group's divergent interests and commitments. At the most abstract level, rhetoricians frame risk in terms of the grand narrative favored by constituents. At the campaign level, they attempt to influence public opinion and to co-opt government regulatory systems. At the ground level, they fight over one contested object after another. In the following sections I'll describe each in turn.

Risk Narratives

A narrative provides a consistent story line through past, present, and future that helps believers situate novel events and respond accordingly. In this section I'll briefly describe three narratives and their respective characterization of transhumanity risks.

"End Times" Narrative

Remember Y2K, the millennium bug? Computer programmers/programs did not take into account the changeover from 1999 to 2000 and as the millennium approached concerns were raised that computer systems, and infrastructures dependent on such systems, would crash. Calculations of risk varied wildly. Andrea Tapia (2003) notes that computer scientists approached Y2K in terms of their experience with patching code. They were confident that they would reduce Y2K to a mere nuisance. In sharp contrast, millennial Christians predicted the collapse of Western civilization. They claimed that a growing dependence on technologies was replacing worship of God and eroding a Christian culture based on faith, community, and simplicity of life. Anticipating a crisis in technological society as a prelude to God's intervention, a Y2K disaster was imagined as an appropriate "punishment from a jealous God." (495).

We find this preoccupation with culture war and the expectation of an epic struggle between religion and the "techno-secular" world in this excerpt from Alan Padgett's article, "God Versus Technology? Science, Secularity, and the Theology of Technology":

> Whether in energy riots or anti-robot revolution, biotechnic warfare or worldwide pollution, or some terrible disaster we cannot now envisage, a totally techno-secular world will eventually destroy itself. Yet, even in this pessimistic scenario, religious faith could provide a small counterculture with an alternative vision that could provide humanity with hope for a future beyond the self-extinction of *homo technicus* (2005).

William Bainbridge (2006:25) envisions a clash between those pursuing "cyber immortality" and religious fundamentalists as a "crucial battle in the long-standing conflict between science and religion."

> Convergenists advocate aggressive research in cognitive science, including computational neuroscience, to understand how the human brain actually creates the mind, and thus how to emulate it…The Convergenists' agenda is aimed at improving human performance without limit, and many of the anticipated technological spin-offs would be useful for recording, preserving, and reanimating human personalities—ultimately creating cyber immortality. Meanwhile, the broad range of valuable new technologies promised by convergence will make the unification of science extremely salient for the ordinary person, thus aggravating the conflict between science and religion. So religion may feel a need to destroy science in order to save itself (28).

Whereas Padgett portrays transhumanity risks in terms of technological disasters, Bainbridge warns that technological success will precipitate civil war. Nevertheless, they both primarily characterize transhumanity risk in terms of social conflict.

Market Exploitation Narrative

Secular conservationists portray transtechnologies within an exploitation narrative that vilifies capitalism and technoscience. Following Martin Heidegger and Jacques Ellu, Leon Kass (2002:35) defines technology as "the disposition to rational mastery." He asserts that commercial interests drive research and development and "soft dehumanization" will occur "unless we redeem ourselves by nontechnological ideas and practices, today both increasingly beleaguered." (22).

Critics of global capitalism, such as Jeremy Rifkin, see transtechnologies extending corporate influence:

> In less than ten years, the global life science companies will hold patents on many of the 30,000 or so genes that make up the human race as well as patents on the cell lines, tissues, and organs of our species, giving them unprecedented power to dictate the terms by which we and future generations live our lives. The concentration of power in the global pharmaceutical industry has already reached staggering proportions. The implications of a new market-driven eugenics are enormous and far reaching. Indeed, commercial eugenics could become the defining social dynamic of the new century (2005:44).

He predicts many negative consequences of a market-driven eugenics including social pressure on families to relent to genetic intervention, a narrowing of the human gene pool, intolerance of those individuals with disabilities, and a de-emphasis on environmental remediation. Mervyn Bendle criticizes Kurzweil's brand of transhumanism:

> Indeed, it is possible to see Kurzweil's futurism as a parable—or indeed as an expression of displaced yearnings or repressed fears—about the dynamics of globalization, because, to a significant extent, humanity is already in a type of posthuman situation. It is posthuman because it has already conceded control of vital areas of human life to the machines and the systems, albeit not (yet) to the forms of cyborg super-intelligence linked in a distributed computer network that Kurzweil envisages, but rather to the supreme super-intelligence of the computer-empowered global market... (2002:61).

Environmentalists see transtechnologies in terms of a larger trend of exploiting nature, including human nature. In an article posted on the Earth First! Journal, Great Grey Owl writes that

> We, as a society, have to get over the idea that private property rights equate to individual liberty and should never be impeded. Until we get over that myth, we will never stop the war machine that is wreaking genocide across the world to suck up resources for "our" interests and "our" sick consumptive dependencies; nor will we stop the mad scientists from bringing about what we long thought were only science fiction horrors—out-of-control robots, genetically engineered monsters, people turned into cyborgs (2006).

Bill McKibben asserts that these technologies follow the logic of consumer capitalism in that "consumption is all that happens, because there's nothing else left that means anything." (2003:46).

For feminists critical of assisted reproductive technologies, for example, Mary O'Brien (1981), Gina Corea (1986) and Robyn Rowland (1992), the backdrop is patriarchy and capitalism and the mistreatment of women in the male-dominated institutions of medicine and science. Laura Woliver (2002) contends that under this dominion "women are wombs who have been skinned by new reproductive technologies so that their innards can be examined, monitored, and possibly controlled by medical professionals and the state" (153). FINRRAGE (Feminist International Network of Resistance to Reproductive and Genetic Engineering) extends the critique to gene technologies:

> The central technique aimed at achieving biological "quality control" today is genetic engineering, a method of analyzing and manipulating the hereditary substance of all life forms. Gene technology is inherently eugenic, because it treats all living beings - micro-organisms, plants, animals and human - as inefficient or outright defective and in need of technical "optimization" to fit the interests of profit and power. Genetic engineering is already being applied to many different areas of our lives - in medicine, agriculture, the industrial production of food, chemicals and other products, by the police and the military. Women will increasingly be faced with the adverse effects, not only with regard to reproduction, but also as producers and consumers, in the areas of food, health care etc. Last but not least, we will all bear the brunt of future ecological disruption, while the profits of the "new genetic revolution" will flow to a few multinationals.

In summary, transhumanity risks are presented in terms of accumulation strategies that are inimical to life. Capitalism is the destructive force and trans-technologies the latest means of exploitation.

New Enlightenment Narrative

Critics of transhumanity are skeptical of reason (instrumental rationality devalues life), progress (a myth used to justify exploitation), and science (effectively under corporate control). Transhumanists, on the other hand, uphold reason, progress, and the virtue of science. Elaine Graham (2003:38) calls transhumanism the "high-tech heir to Enlightenment humanism" and I believe that is a fair characterization in that transhumanists cherish these Enlightenment values. They firmly believe in the Enlightenment project of using reason to improve the human condition. Progress is unabashedly proclaimed and, as Simon Young states, the "new technologies are joyously celebrated as the wonders of the modern world." (2006:20) Gregory Stock evokes Benjamin Franklin's enthusiasm for industry, innovation, and science. He favors daring over hand wringing.

> As we push further into uncharted territory by deciphering and laying bare the workings of life, it is worth asking just what it is that so worries us. The enormity of coming developments in molecular biology seems obvious, but their magnitude does not require that we

respond with fear. We are hardly the first to appreciate the prospects for impending revolutionary developments in science. In 1780, Benjamin Franklin showed a very different attitude, when he wrote to the great English chemist Joseph Priestley, "The rapid progress true science makes occasions my regretting sometimes that I was born so soon. It is impossible to imagine. the heights to which may be carried, in a thousand years, the power of man over matter." (2003:679).

James Hughes accuses "Left Luddites" of failing to follow this spirit.

Most political radicals embraced science and technology in the eighteenth and nineteenth century. But another group of radical critics have seen technology as an intrinsic part of the evils of capitalism, patriarchy, and racism. Since World War Two techno-skepticism has come to replace techno-optimism on the Left (2003:125).

Risk is not denied but it is placed within the context of modernity's successes, for example, reduced infant mortality, increased longevity, long distance communication, and leisure time. Nothing ventured, nothing gained. Rather than emphasize alienation and victimization, the transhumanists exude confidence in a proactive approach to sociotechnical ensembles. Andy Clark states it this way:

And we do need to be cautious, for to recognize the deeply transformative nature of our biotechnological unions is at once to see that not all such unions will be for the better. But if I am right, if it is our basic human nature to annex, exploit, and incorporate nonbiological stuff deep into our mental profiles, then the question is not whether we go that route, but in what ways we actively sculpt and shape it. By seeing ourselves as we truly are, we increase the chances that our future biotechnological unions will be good ones (2003:198).

Here we see again the transhumanist belief that human nature is about innovation and transformation. Risk-taking is what we do as a species and transhumanity risk-taking is no different. This is so remarkably different from the religious conservationists' warning that transhumanity risk-taking will be cataclysmic or the secular conservationists' charge that it will allow for the expansion of an exploitive system.

Risk Campaigns

The key objectives of a risk characterization campaign are to mobilize the movement base, influence public opinion, and enlist the support of legislators in an effort to maintain or change the regulatory system. Specifically, transhumanists and their opponents compete to shape 1) perception of whom to trust, 2) the overarching management principle of the oversight system, either precautionary or proactionary, and 3) liability, i.e., laws regarding culpability and compensatory or punitive damages for injury.

Trust

Often there are surplus claims regarding risk. How can we decide which claim to believe? We can turn to the evidence, if it is provided, but even there we may face the problem of contradictory evidence. Furthermore, as Harry Collins and Trevor Pinch (1993:98) conceptualize in the "experimenter's regress," the methods used to produce evidence can always be challenged by an opposing party. With the transhumanity debate we face the additional difficulty that the hazards of trans-technologies are unknown.

At some point as we wrestle with competing claims regarding risk, we may move beyond a consideration of the case-as-presented and consider instead the trustworthiness of the case presenter. Whom do we believe? What authority do we recognize? For example, for some consumers a regulatory agency's approval of a prescription drug is sufficient authorization to accept a pharmaceutical company's claims for the drug, but not for those consumers mistrustful of the agency or company.

Ad hominem attacks, especially to impugn motives, are designed to sow distrust. Both sides in the transhumanity debate make use of such attacks. The transhumanists label their adversaries "fundamentalists" and charge that they are engaged in a broader assault on modernity. Ray Kurzweil (2005:415) likens environmentalist "Luddites" to Islamic extremists in that both groups wish to impede or repeal progress. He asserts that for relinquishment to work it would have to be "totalitarian relinquishment," a massive social control system that would effectively end democracy. (2005:407) Ronald Bailey (2005: 218) argues that the regulatory process should be based on "objective scientific standards" and not "societal values" because the latter would serve as a Trojan horse for activists' political agenda, for example, opposition to free trade and globalization. Gregory Stock (2002:154) castigates a committee under the auspices of Association for the Advancement of Science for including "theologically acceptable" as a condition for inheritable genetic modifications. He sees this as an incursion by religion into science's domain and a power grab by the religious right. The transhumanists' overall message is that you can't trust the "bioconservatives" or "fundamentalists" to tread lightly when it comes to science policy. After scaring the public into ceding them new authority, the fundamentalists will impose a reactionary political agenda.

There is more variation in the conservationists' ad hominem attacks, reflecting the diversity of interests, but a common tactic is to question the motives of scientists and engineers. Scientific authority is reinforced by a flattering portrait of scientists/engineers as dedicated and selfless men and women intent on discovering the truth about nature or creating products that are used to improve the human condition. This noble figure, however, has an alter ego in Western literature—the mad scientist. Mary Shelley's Dr. Frankenstein, for example, breaks moral laws and mistreats others in an all-consuming desire to achieve the God-like power of reanimation. In his sweeping account of Christian millennialism and its effect on the Western technological enterprise, David Noble criticizes many legendary scientists for pursuing transcendence regardless of the social consequences. He charges that in modern times scientists have financed their research by selling out to military and corporate interests:

> As they have trained their minds for transcendence, they have contributed enormously to the world arsenal for warfare, surveillance, and control. And they also have placed their technological means at the disposal of manufacturing, financial, and service corporations, which have deployed them the world over to discipline, deskill, and displace untold millions of people, while concentrating global power and wealth into fewer and fewer hands" (1999:206).

Others emphasize personal gain. Leon Kass, for instance, asserts that for "some scientists and biotechnologists" and their entrepreneurial backers "[t]here are dreams to be realized, powers to be exercised, honors to be won and money—big money—to be made" (2002:6). According to Kass, ambition clouds judgment to the extent that scientists and engineers fail to recognize when their experiments and creations put others in jeopardy.

Transhumanists take offense at this characterization, just as the conservationists take offense by being described as fanatics. The contestants assert, correctly I believe, that the caricature does not do them justice. Although ad hominem attacks may resonate with one's base, in my opinion they are ultimately corrosive to the debate.

Oversight Based on the Precautionary Principle

Urlich Beck (1992:19) asserts that advanced modernity is characterized by "the social production of risks" in at least two ways: 1) the economic production of hazards such as pollutants and toxins, and 2) sociopolitical enterprise built around the promise of security and "discovering, administering, acknowledging, avoiding or concealing" risks (20). Beck criticizes this oversight system for being too reactive and permissive of the production factors that generate hazards. Most importantly, burden of proof is placed on consumers or advocacy groups to precisely identify hazards, establish cause and effect, and calculate future risks—a task that Beck (1995) asserts is made nearly impossible by the complexities of global production.

Furthermore, risk assessments are susceptible to contestation in a number of ways. Surveys and interviews can be criticized for their limited explanatory power whereas other methodologies including animal modeling, computer simulations, and controlled field trials may be contested on the grounds of poor validity or lack of fidelity to real world conditions. Predicting how a new technology will be used or misused, the complex interactions, and far reaching consequences is inherently difficult and open to attack. Moreover, as R.J. Berry (2006) notes, deciding the *acceptability* of risk entails ethical judgment, and ethical argumentation can be challenged by the application of a different ethical theory or an appeal to a different set of values. In conclusion, the party that must prove its risk case is at a distinct disadvantage.

Beck calls for a redistribution of the burden of proof. Environmentalists' precautionary principle would do just that. According to this risk management scheme, producers would have to demonstrate short term and long term safety of products or systems before deployment. The Wingspread Consensus Statement on the Precautionary Principle (1998) calls for caution as the default response to risk uncertainty:

> When an activity raises threats of harm to human health or the environment, precautionary measures should be taken even if some cause and effect relationships are not fully established scientifically. In this context the proponent of an activity, rather than the public, should bear the burden of proof.

If the precautionary principle were adopted as the guiding principle for the oversight of transtechnologies, the individuals and corporations that are advocating transtechnologies would have to swim against the tide. This line of attack may succeed where faith-based campaigns fail. For instance, conservative Christians insist that human life begins at conception and that the deliberate cessation of human life at any period during gestation is an immoral act. They have condemned embryonic stem cell research (ESCR) and related transtechnologies in terms of this stand on abortion. Fellow believers are persuaded by these arguments, but what about non-believers? Can a case built on religious doctrine succeed in pluralist societies with rational-legal institutions? In contrast, the precautionary principle works within the secular system and, if adopted, it would have the secular system work for conservation. We get a sense of what a precautionary oversight system would look like from George Annas:

> We need to exercise our moral imaginations to create a structure that can act as a virtual global conscience for the scientific community pursuing species-altering and potentially species-endangering biotechnologies. An ethical oversight structure must be global and should include representatives from governments, industry, non-governmental organizations and the public. The group should be charged with articulating substantive global research rules (using existing international human-rights documents, like the Nuremberg Code, the Covenant on Civil and Political Rights, and the European Convention on Human Rights and Biomedicine as a basis), reviewing and approving (or declining) all proposals to do species-altering or potentially species-endangering procedures, and monitoring these experiments as they are performed (2006).

Oversight Based on the Proactionary Principle

Transhumanists are well aware of the implications of the precautionary principle. Ronald Bailey points out that "accurately predicting in advance the benefits and harms that a technology may one day produce is impossible. This inherent uncertainty means that opponents of a new technology can always stall its introduction by endlessly demanding more research." (2005:213) Understandably, he favors the current regulatory system which is not so demanding. James Hughes and Gregory Stock also prefer the status quo over a precautionary system, insisting that the latter would stifle technological innovation and medical breakthroughs.

Max More and others who participated in the Extropy Institute's Vital Progress Summit I in 2004 took up the challenge presented by the precautionary principle, and from this meeting More (2005) crafted the proactionary principle as an alternative risk management scheme. Whereas advocates of the precautionary principle have identified biases in prevailing risk management systems, More (2005) alleges the following biases in the precautionary framework: assumes worst-case scenarios, fails to take into account the negative consequences of restriction, favors nature over humanity, and is "skewed against economic and technological progress and development."

For every precautionary claim More offers a proactionary counterclaim. Precautionary claim: New technologies pose the greatest risk to humanity. Proactionary counterclaim: New technologies provide the means to handle grave threats such as food shortages, natural pathogens, and environmental changes. Precautionary claim: Priority must be placed on preventing harm. Proactionary counterclaim: Costs *and benefits* must be taken into account and one should "estimate the opportunities lost by abandoning a technology, and take into account the costs and risks of substituting other credible options." Precautionary claim: Even if it is determined that the probability of negative consequences is low, it is prudent to relinquish a technology that might cause serious harm. Proactionary claim: "Consider restrictive measures only if the potential impact of an activity has both significant probability and severity." It is preferable to handle negative effects through "compensation and remediation instead of prohibition."

Assignment of Liability

Risk can be conceptualized as being spread out—a system characteristic—or it can be pinned down to a particular source. For example, the risk of highway accidents can be attributed to the constellation of drivers, automobiles, roads, and weather. Injuries and fatalities may be perceived as an unfortunate price to be paid for having a mass transportation system. In contrast to this system approach, one element may be singled out for blame. If drivers are deemed the responsible party, risk might be managed by means of more stringent education requirements, surveillance, and penalties. If automobiles are so designated, manufacturers might face greater scrutiny and regulation. Wetmore (2004) describes how consumer advocacy groups and insurance companies in the past few decades have been able to shift more responsibility unto car manufacturers. The installation of air bags in all new vehicles is one tangible result. In many cases there are no commonly accepted, definitive methods to localize blame and, accordingly, interest groups compete on this matter. Campaigns can be waged for decades.

Transhumanists utilize a systems approach when they describe the risks associated with new technologies as a necessary condition of progress. Max More (2005) asserts that the advancement of civilization could not have happened without taking risk:

> If the precautionary principle had been widely applied in the past, technological and cultural progress would have ground to a halt. Human suffering would have persisted without relief, and life would have remained poor, nasty, brutish, and short: No chlorination and no pathogen-free water; no electricity generation or transmission; no X-rays; no travel beyond the range of walking.

He readily admits that transhumanity, too, entails risk, however the expected gains in life expectancy, vitality, creativity, and comfort are worth the price. Gregory Stock argues that there is no way around trial and error—we "have to earn our knowledge by using the technology and learning from the problems that arise" (11).

When the transhumanists are drawn into the dispute over localizing blame they tend to uphold the status quo. They argue that manufacturers of transtechnologies should be held accountable for their products but according to the same standard of liability faced today by the medical and pharmaceutical industries. Manufacturers are liable for damages if the consumer suffers physical injury and suffering primarily if risks were not disclosed, the safety testing was inadequate or falsified, or the manufacturer failed to respond to reports of harm. The law often allows exposure to risk (for example, drug side effects) and it does not recognize alleged damages to human nature, the social system, or religious life. Transhumanists want to keep it that way.

Transhumanists also hold consumers responsible for their purchasing decisions. They should carefully evaluate products and weigh the costs and benefits. Gregory Stock (2002) trusts consumers' discernment and he believes that they will select safe products over harmful ones. He also trusts parents to act responsibly when choosing therapies or enhancements for their children.

Conservationists contend that there are unprecedented risks associated with transtechnologies and that they should be subject to a much more stringent safety testing and liability system. Bill Joy (2000) warns that genetic and nano engineering pose a new danger to humanity in that self-replicating microorganisms or nanobots might breach containment and run amok. Even if the probability of catastrophe is low, it is unacceptable to allow experimentation to proceed.

The *success* of transtechnologies, as well, may produce negative, unintended consequences. Leon Kass (2002) worries that consumers may opt for neuro-transformations that, like the action of psychotropics, reduce anguish and promote pleasure but also diminish character development and life purpose. Whereas transhumanists see no ethical or legal problems with a voluntary system of self-transformation and emphasize the benefits to health, longevity, and ability, Jeremy Rifkin (2005) asserts that those resisting market driven augmentations will face higher insurance premiums or coverage denial and job discrimination. The human community may suffer a schism with extreme social inequality between traditionalists and those swept up in the spiral of social competition/augmentation. According to Francis Fukuyama, a "genetics arms race will impose special burdens on people who for religious or other reasons do not want their children genetically altered." (2002:97) These critics argue that social damages need to be taken into account and they justify banning enhancement technologies in order to prevent social conflict.

George Annas alleges that scientists involved in research on xenografts (transplantation of nonhuman tissues or organs into human recipients) and germline genetic engineering place the human species in jeopardy. He would have them liable for crimes against humanity.

> No individual scientist or corporation has the social or moral warrant to alter or endanger the human species. We recognize this most readily when scientists use their expertise to develop a new bioweapon, or a new pathogen that can more efficiently infect or kill humans, and we can reasonably term such scientists bioterrorists.
>
> Scientists who use novel techniques to try to improve the lot of humans are not, of course, in the same category as terrorists. Nonetheless, when scientists engage in species-endangering experiments in the absence of a social warrant, and ignore the human-rights implications of their work, their activities can be considered a terrorist act. Some xenografts, for example, carry the risk of releasing a new, lethal virus on humanity. Germline genetic engineering likewise poses threats to the human species by inevitably moving Homo sapiens to develop into two or more separable species: the standard issue humans and the genetically enhanced posthumans; the former will likely be seen as savages and heathens that the posthumans can properly slaughter and subjugate. It is this genocidal potential that makes some species-altering genetic engineering projects potential species-endangering weapons of mass destruction, and the unaccountable genetic engineer a potential bioterrorist (2006).

Transhumanists do not accept that human genocide is inevitable, for instance, James Hughes (2004:261–262) envisions future democratic societies that are hospitable to humans and transhumans by "expanding the bounds of tolerance and equality" and guaranteeing the "right of all persons to control their own body and mind."

Transhumanists oppose excessive state regulation and red tape on the grounds that it stifles innovation, nevertheless, their approach should not be characterized as laissez-faire. They accept the state's role in controlling technologies that present a clear danger to the populace or environment. For instance, Ray Kurzweil recommends a prohibition on self-replicating nanobots. Others worry about runaway artificial intelligence and would support a moratorium on research if warranted. Although government oversight systems are criticized for being ponderous, most transhumanists accept the status quo and pursue a defensive strategy to prevent opponents from co-opting the oversight system and adding new mandates that would effectively stymie transhuman technologies. James Hughes is very candid about this:

> The bioLuddites are unhappy with the idea of transhuman technologies being under existing agencies, such as the U.S. Food and Drug Administration, which are mandated only to ensure safety and efficacy. Instead they want new agencies empowered to ban technologies on the grounds of vague long-term risks such as the future conflict between

humans and posthumans. Consequently transhumanists should promote the use of the existing agencies to regulate transhuman technologies in order to protect more liberal access (2004:241).

Contested Objects

To gain acceptance of a scientific theory or market an invention it helps to be resourceful. Peer support and funding are just as important as the brilliant idea. John Law coined the term, heterogeneous engineering, to emphasize that entrepreneurial engineers must tackle economic, political, and social problems, as well as, technical problems to be successful. Thomas Edison, for example, was not only a brilliant inventor, he also was a shrewd businessman and adroit at manipulating the US patent system. In the 1980s and early 1990s actor network theorists, most notably Bruno Latour, Michel Callon, and John Law, highlighted the importance of alliance building for waging technoscientific campaigns. They pointed out that humans and non-humans (for example, anthrax bacillus, lab equipment, log books, photographs) serve as allies. Susan Leigh Star and James Griesemer (1989:508) noted that "boundary objects" are particularly useful because these objects inhabit several intersecting social worlds" and can be used as points of translation and coordination. For instance, a medical website could serve as a boundary object through which a physician might secure a patient's acceptance of a particular treatment. However, alignment of interests is not inevitable and boundary objects can become points of contention. If new postings on the website are critical of the treatment, the patient may opt out.

Transhumanists and their detractors use high visibility boundary objects as a way to capture public support. I will describe a subset of those objects that 1) have a high risk profile and 2) are strenuously contested by both sides. The "risk objects" that I will discuss here are terrorism, genetically modified (GM) food, and neuropharmaceuticals.

GNR Terrorism

The costs associated with terrorism are quite high in terms of deaths, injuries, damage to buildings and infrastructure, the long term expenditures in providing security and waging war, and the emotional price of living in fear. Few phenomena in the new millennium match terrorism for real and symbolic import, and this makes terrorism a very valuable boundary object. Politicians make use of this object in political debate, for instance to justify suppression of independence movements or to smear opponents ("He's weak on terrorism.") Some antagonists in the transhumanity controversy are not above using it in ad hominem attacks, for example, referring to stem cell researchers as "terrorists of the unborn".

A key contested issue in the transhumanity debate is whether terrorists will use transtechnologies as weapons of mass destruction. It might be helpful to refer once again to Bill Joy's argument regarding the peril of genetic, nano, and robotic (GNR) technologies. In contrast to nuclear, biological and chemical (NBC) weapons, Joy (2000) describes GNR technologies as being

> widely within the reach of individuals or small groups. They will not require large facilities or rare raw materials. Knowledge alone will enable the use of them. Thus we have the possibility not just of weapons of mass destruction but of knowledge-enabled mass destruction (KMD), this destructiveness hugely amplified by the power of self-replication.

Also, unlike NBC weapon systems that have been developed under military control, these new technologies are being developed by capitalist enterprises that follow an imperative to disseminate their products. Joao Pedro de Magalhaes (2002:42) assumes easy access to these technologies in the near future and warns that

> Eventually, we can reach the point where a single human will have the power to kill all other humans. Since among the billions of humans alive today we can find many willing to destroy humankind, human civilization would end. A civilization where anyone has the power to destroy everyone cannot stand. Even before reaching such a point, we are likely to suffer major catastrophes as a result of terrorist attacks using weapons of mass destruction. If every individual has access to knowledge and tools capable of mass destruction, our future will be a future of death and destruction.

Conservationists insist that the only way to prevent this is to keep Pandora's box closed and to follow Joy's recommendation of relinquishment. That would entail a comprehensive moratorium on all research in these areas. The ostensible purpose of relinquishment is to deny terrorists such dangerous tools, but of course it would also serve the conservationists' goal of stopping transhumanity. By characterizing transtechnologies as potential tools of terrorists, they become tainted by association. Furthermore, it becomes easier to make the case that scientists and engineers developing GNR technologies are placing their own professional ambitions above the common good. The transhumanity project, with its lofty aspirations, becomes weighted down.

Transhumanists could charge that their adversaries exaggerate, however this might appear inconsistent with their own claims in the power of these technologies. Instead they utilize an interesting counter strategy of embracing the risk object. They assert that GNR research is taking place around the globe and that there is no way to return to a period of innocence. Even if nations agreed to relinquishment, clandestine operations would continue. Gregory Stock (2002) warns that rogue regimes or terrorists would develop weapons against which peaceful nations would be helpless to defend. The risk of mass destruction would increase, rather than decrease. The safest policy for democracies is to always stay ahead in the advancement of technologies in order to provide effective countermeasures.

Ray Kurzweil (2005:412) proposes "fine-grained relinquishment" (412) as an alternative to sweeping moratoriums as it is flexible enough to allow for medical advances, target for suppression the most serious threats (e.g., self-replicating nanobots), and provide GNR defensive measures:

> Technology will remain a double-edged sword. It represents vast power to be used for all humankind's purposes. GNR will provide the means to overcome age-old problems of illness and poverty, but it will also empower destructive ideologies. We have no choice but to strengthen our defenses while we apply these quickening technologies to advance our human values... (424)

In his "Program for GNR Defense," (2005:422–424) Kurzweil calls for defensive technologies developed through GNR research, some to be deployed in the environment to serve, for instance, as early-warning systems, and others to be embedded in the body to boost immunity. For Kurzweil the way out of terrorist threats is to go forward with transhuman enhancement. The risk object is appropriated.

Genetically Modified Food

The production and distribution of genetically modified (GM) food is controversial, with environmentalists pitted against agribusiness, organic farmers against those that cultivate GM crops, and exporting nations such as the United States, Argentina, and Brazil clashing with the European Union. Through genetic engineering, desired properties, for example, resistance to cold, drought, rot, and pests, are introduced into a plant. Proponents extol benefits over conventional crops: increased yields, reduced use of pesticides, decreased water consumption, and higher levels of nutrients. Opponents emphasize costs: allergenicity, reduction of crop diversity, harm to beneficial insects, pest resistance, and escape into the wild.

Proponents argue that social equality is promoted to the extent that GM crops bolster the world supply of food, local farmers obtain higher yields and spend less on pesticides, and vitamin deficiencies are reduced. Others charge that social inequality increases from the economies of scale favoring industrial agriculture and higher price of GM seeds. Organic farming may suffer crop contamination by cross-pollination with GM products. Natural insecticides such as Bacillus thuringiensis (Bt) used by organic farmers may lose their effectiveness (when introduced into GM plants, insects may develop resistance over time).

Transtechnologies are still many years in the making but in 2009 GM crops were raised on approximately 330 million acres (Clive 2009) and GM food is sold in grocery stores around the world. Also, media attention has helped raise awareness with one poll finding 40 % of American respondents having come across information (Mellman Group 2006) and only 26 % of European respondents indicating that they lacked information (Cage 2008). All this makes GM food an excellent boundary object and proxy for the transhumanity dispute.

Granted, there is no direct link between GM food and transhuman enhancement. Detractors question whether so-called Frankenfood is safe for human consumption, but few would assert that people are fundamentally altered. Transhumanists and conservationists, instead, emphasize associations between the two—both involve high-tech, entail the transformation of species, and are justified according to a utilitarian argument.

The dispute unfolds just as one would expect with the transhumanists highlighting the alleged positive consequences while the conservationists emphasize the negative. Neither side is willing to characterize it as a mixed bag. Transhumanists portray conservationists as Luddites, alarmists, and hypocrites for coming across as champions of the poor but opposing the technologies that will best keep the poor fed and healthy. Conservationists depict transhumanists as being dupes or lackeys of biotech corporations, unwilling to take into account ill effects, and insincere in their concern for the poor. Bill McKibben (2004) sees GM crops as the latest episode in a history of exploitative food policy.

> For one thing, the whole history of U.S. "food aid" to poor countries has been considerably less noble than the words would imply. Not only has food been used as a political bargaining chip on countless occasions, it has also played a key role in wrecking the agricultural infrastructure of one country after another. As our subsidized cheap corn and wheat flood in, local farmers can't compete...
>
> And with the advent of genetic engineering, our biotech corporations have started using that same kind of power to force GM crops onto poor countries and hence into the world foodstream. The Europeans won't buy our corn, but perhaps it can be rammed down African throats, and into their fields, and hence make its spread a fait accompli.

According to the conservationists, if we can't trust biotechnology corporations to genetically engineer food without jeopardizing health, intensifying social stratification, and threatening the natural system, why should we trust them with our bodies?

Change can be defined as radical, hence more risky, or normal and within our ability to anticipate and control. Conservationists describe gene modification in plants as being discontinuous with previous practices and, accordingly, presenting unknown dangers—an argument that they extend to human genetic engineering. Transhumanists counter that genetic modification of plants is not significantly different from traditional breeding practices. Humans have for thousands of years selected and controlled for plant genes. Ronald Bailey states that "[f]arming, it's worth remembering, is the opposite of letting nature run wild." (211) He contends that the new techniques allow for more versatility. He also asserts that they are being improved in order to reduce unintended consequences such as allergenicity. In his view, these developments are in line with progress. Furthermore, the knowledge and skill sets gained from genetic engineering of plants and animals provide a foundation for human research.

Neuropharmaceuticals

Prozac and Ritalin are just a few of the many drugs widely prescribed to improve emotional state, disposition, and functioning. Over the last decade prescriptions have increased. In the United Sates the number of antidepressants drugs recorded during visits (physician and hospital outpatient) from 2004-2005 per 100 population was 35.5 and for those under 18 years of age the amount was 8.9 for anorexiants/CNS stimulants (to treat attention deficit disorder, hyperactivity). The figures for 1995–1996 were much lower, 13.8 and 3.9, respectively. (National Center for Health Statistics, 2007:333) Although drug treatment has its private, confidential side in the context of the patient-physician relationship, multi-million dollar advertisement campaigns and a spirited anti-medicalization backlash have amplified visibility, thereby making neuropharmaceuticals a good boundary object over which to fight. Once again, the adversaries tend to stake out diametrically opposed positions with the transhumanists happy to cite neuropharmacology as a case of successful technoscientific intervention while the conservationists make it the subject of a cautionary tale against enhancement.

The transhumanists emphasize the therapeutic value of these drugs and give credit to drug makers and physicians for acting with beneficence as they help alleviate their patients' pain and suffering. They select favorable testimonials by users describing positive life change—from social withdrawal to participation, incapacitation to productivity, and despair to hope. Always thinking one step ahead, they imagine that new drugs or other neurotechnologies will not only target pathologies but extend the upper range of emotional and cognitive functioning. For instance, David Pearce (1998) foresees a time when nanotechnology and genetic technology will be used to promote a higher state of well being:

> Over the next thousand years or so, the biological substrates of suffering will be eradicated completely. "Physical" and "mental" pain alike are destined to disappear into evolutionary history. The biochemistry of everyday discontents will be genetically phased out too. Malaise will be replaced by the biochemistry of bliss....
>
> This feeling of absolute well-being will surpass anything contemporary human neurochemistry can imagine, let alone sustain. The story gets better. Post-human states of magical joy will be biologically refined, multiplied and intensified indefinitely. Notions of what now passes for tolerably good mental health are likely to be superseded. They will be written off as mood-congruent pathologies of the primordial Darwinian psyche.

Conservationists, in contrast, emphasize the risks involved in drug use and argue that these risks will be amplified with more powerful neuro-manipulators. Francis Fukuyama's (2002) critique is representative. Treatment of mental illness is appropriate but prescription drugs should not be used for mood enhancement or social control (e.g., to manage children). Individuals must accept some discomfort and pain that comes along with personal development. Prescription drugs are not an unalloyed good, rather they amount to a devil's bargain: "freedom from depression, together with freedom from creativity or spirit; therapies that blur the line between what we achieve on our own and what we achieve because of the levels of various chemicals in our brains" (8).

Fukuyama worries that multinational pharmaceutical corporations with a financial incentive to expand their product lines will continue to medicalize relatively minor personal troubles and that "the psychiatric profession could probably be depended on to declare unhappiness a pathology" (56). Of course, many people today use pharmaceuticals for non-therapeutic reasons. According to the International Narcotics Control Board (2006), prescription drug abuse has increased in recent years. For example, "the number of Americans who misused controlled prescription drugs nearly doubled from 7.8 million in 1992 to 15.1 million in 2003." (Zarocostas 2007) Even if negative side effects and overdose could be eliminated, Fukuyama cautions the "perfect" drug that provides happiness or pleasure would come at the cost of dependency, complacency and submissiveness.

Protecting the "Risk Object Portfolio"

The transhumanists invariably highlight value added by sociotechnical transformations whereas conservationists tend to emphasize value lost. We find this pattern across a variety of risk objects, not just the ones discussed above, but also assisted reproductive technologies, life extension technologies, and virtual reality. Consistency of *message* is most important and, as with military campaigns, alignment of forces and a unified front are crucial. Another way to think of this process is in terms of managing investment portfolios. By pronouncing allegiance or opposition to risk objects, the contestants thereby stake an interest in the object and its future. Both sides are committed to their respective portfolio, so they must be prepared to respond with flexibility and improvisation to "market developments" over which they have little or no control. Contestants will continue to exercise "spin" to emphasize or deemphasize certain developments.

Conclusion

For the time being I expect transhumanists and conservationists to characterize risk in a way consistent with their respective narrative and in accordance with the views of their respective base. They will continue to use risk objects as proxies for transtechnologies. They will continue to drum up public support and pressure legislators, but I doubt the regulatory system will change dramatically one way or another while transtechologies are years away from development. There will remain many degrees of freedom in this phase. I expect this to change when and if transtechnologies face clinical trials. Documented success or failure, both medical and social, will place constraints on strategies and tactics of risk characterization. Facts will emerge regarding costs and benefits and, although these may be contested, they will be given much more weight politically thereby providing more leverage for one side or the other to influence the oversight and regulatory systems.

Transhumanists are so confident that transtechnologies will fare well and will eventually become diffused in society that they depict transhumanity as being inevitable. This can be taken for bluster but I see it as another rhetorical strategy crucial during this preemption phase. Those uneasy about transhumanity may not be persuaded by the transhumanists' risk rhetoric, but they may be dissuaded from active opposition if they believe that transhumanity will happen anyway. The conservationists must convince the public that relinquishment is the prudent course and that it is possible to halt socio-technological change. I'll discuss this and other important aspects of the dispute over inevitability in the next chapter.

Chapter 5
Inevitability

Looking out the window on this summer day my sight is drawn to the contrasting yellow and purple flowers in the garden. The plants sway gently in the light breeze. Bees move from flower to flower. My attention shifts when Sucha, a beautiful Siamese cat, leaps onto my desk and settles on her folded blanket. Her eyes are a deep blue, her face and ears strikingly black. The house is quite…

We all have these ordinary moments when we are effortlessly involved in the present. I gained an appreciation for these moments when I took up meditation a few years ago. Being a novice I have a better theoretical understanding than working knowledge (typical for a professor). I've learned that meditation requires a quiet state and release from preoccupations about the past and future. The disagreement with a colleague—let it go. Nagging worries about making the children's tuition payments in the next few years—put them aside. Try instead to be conscious of the here-and-now, and, if it helps, focus on breathing, repeat a phrase, or become aware of the surrounding environment. I find it helps if I entertain the thought that all my living takes place only in the present. The past is gone and unreachable and I cannot yet experience tomorrow, a year, or a decade hence. The before and after do not have any bearing on my present state.

I want to make clear that presentism is not my everyday attitude toward time. Frankly, I am not sure whether I could live my life without reference to the past and future. I plan and act with time trajectories in mind: "My daughter was too young last year to drive a car, but now that she has passed the learner's permit test I'll have to set aside time on the weekends to help her practice." "I've completed four chapters of the book and I hope to finish this chapter before the semester begins. Can I complete the book by this Fall?"

Most people treat the past, present, and future as connected. The contestants in the transhumanity debate are no different; they just go public with their account. They identify trends from past to present and project them into the future. Besides disagreements over what constitutes optimal progression, the most contentious issue is whether a particular future is inevitable.

S. Lilley, *Transhumanism and Society*, SpringerBriefs in Philosophy,
DOI: 10.1007/978-94-007-4981-8_5, © The Author(s) 2013

Rhetoric of Inevitability

Short of some breakthrough in physics involving time travel that all sides would accept as providing rock solid evidence of the future, there is no way to prove inevitability. We can't even be certain that we will live another day. Powell (2000) describes 20 scenarios of how the world could end suddenly, e.g., asteroid impact, gamma ray burst, collapse of the vacuum, super solar flare, and particle accelerator mishap. Still, most of don't expect the sky to fall. We assume that life will carry on. Moreover, we don't demand proof for whatever version of the future that we come to accept.

Acceptance of an inevitability claim has implications for behavior and collective action in accordance with the dictum by W.I. Thomas: "If men define situations as real, they are real in their consequences." For instance, European leaders in 1914 were so convinced of the inevitability of warfare (due to binding alliances, armaments build-up, ethnic strife, etc.) that it became a collective-fulfilling prophecy in the reality of World War I. Since World War II, billions of humans have been subjected to another experiment in the psychology of inevitability: the centerpiece of nuclear war deterrence– MAD (mutual assured destruction)—is based on the premise that leaders will not launch a nuclear strike as long as they are convinced of the inevitability of retaliation and their nation's ruin.

Transhumanity and Fatalism

I am sure that the transhumanists would prefer that individuals embrace transhumanity because they are inspired by its prospects, but resignation will do if it means a lack of resistance. A few years ago I conducted a qualitative survey of college students to obtain a sense of how young adults perceive transhumanity (Lilley 2007). Are they thrilled or frightened? Do they see potential benefits or threats to their way of life? I found that nearly 3 out of 4 respondents held negative attitudes toward transhumanity. Over 90 % of the students indicated that religion would suffer, and many presented this as a personal matter. For instance, one student explained that

> Without the respect in God's power to create the beautiful complex humans we are, and without the belief in prayer, or faith, there is no basis of religion. In a transhumanistic society, there is much less emphasis on beliefs, spirituality, and the power of a higher being, but more emphasis on the power of science and technology. As a believer of God, I support the belief that God created earth, man, and everything in it…Without true human beings, who are created in the likeness of God, we would be forced to abolish all hope for an afterlife, such as Heaven.

Many of the students discussed what can be done, if anything, to influence transhumanity (including preventing it) and more than twice the number expressed resignation as they did opposition. One respondent described acceptance as the most prudent course:

Evolution in every aspect is inevitable... If it wasn't for new technology and medical research we wouldn't be as advanced as our present day culture is. The world and technology are improving as the day goes on. Everyone has to accept that in order to move on with life, one needs to accept that this is the way things are going to be. Transhuman will pop up in the future more and more.

Another respondent conveyed a more worried fatalism: "It is impossible to stop coming up with new technologies and advancing. I just wonder sometimes if we will ever cross the line."

I was astonished by such responses for it appears that many students have accepted inevitability claims, and these seem to blunt opposition to transhumanity. If true for the larger population, this would have profound implications for the future. To reiterate, there is a lot at stake over the rhetoric of inevitability. The contestants know this and claims and counterclaims abound. Let's take a closer look at this controversy.

Strong Claims of Inevitability

Whereas the conservationists often take the offensive in the rhetoric of risk, the transhumanists clearly are the aggressors in this contest. First, the transhumanists convey a *sense* of inevitability through their sweeping account of technological innovation. This is most effective when describing human history in terms of successive waves of beneficial technologies used to alter, control, or bypass nature, for example, fire-building, agriculture, vitamins, and vaccines. According to proponents, transtechnologies represent the next step in progress.

Second, outright inevitability claims are made. I'll categorize such claims as "strong" when transhumanists declare an immutable process and "moderate" when they assert that social conditions are, and will continue to be, favorable but not necessarily causal for the development of transhumanity. I'll start with the strong claims by identifying the immutable process and where is it located: (1) Evolution—it operates either in the broad expanse of the universe or, more specifically, life on earth. (2) *Homo cyberneticus* and the drive to self transform— it is deeply ingrained in human nature. (3) Technological momentum—exponential growth thought to be an internal dynamic of technologies. Transhumanists often depict these processes overlapping, however for the sake of clarity I will describe each in turn.

Evolution

In the most general sense, evolve means to develop over time. When used to describe change in organisms over generations the connotation of improvement often becomes pronounced (especially when our species is the subject). Charles

Darwin preferred the phrase *descent with modification* rather than *evolve* because the former was neutral regarding adaptations. A modification might provide functional advantage to an organism but always in response to the demands of the environment at a particular point in time. Environments change too, and a subsequent modification is "judged" in relation to its time and place. As Stephen Jay Gould asserts, it is all about "local adaptation, not of general advance or progress." (2007, p. 209)

It is often suggested that the concept of evolution dealt a blow to the conceit that, of all living beings, humans are unique and privileged. However, this depends on what version we accept. If we follow Gould's approach it is hard not to be humbled. He argues that evolution is a messy affair: punctuated and not gradual, affected by cataclysms (e.g., asteroid strike), and very complex–competition or symbiosis playing out at many levels (genes, individual organisms, and collectives of organisms). Human evolution and human consciousness were not preordained: "Humans arose, rather, as a fortuitous and contingent outcome of thousands of linked events, any one of which could have occurred differently and sent history on an alternative pathway that would not have led to consciousness." (211) By asserting that bacteria are the champions in terms of adaptability and long term success, he challenges the assumption that humans' most highly regarded faculties make our species the best fit.

Gould's take on human evolution is not shared by the general public. The common understanding is that humans are the most evolved species with regard to thinking, language, and sociality. For many it is a source of pride to think that nature selects for these capacities and our species has come out on top. Transhumanists tend to utilize this flattering interpretation but add the caveat that the selection process continues and that humans will not be the pinnacle of evolution. Kurzweil (2005) argues that intelligence provides a competitive advantage because "[i]ntelligence, if sufficiently advanced, is, well, smart enough to anticipate and overcome any obstacles that stand in its path." (206) Modifications that increase computation power tend to be retained, and in the long run the trajectory is ever upward. Humans are on the high end of the continuum of smart species but we will evolve in synergy with our most advanced technology to become higher-order computation beings.

In his book, *Mind Children*, Moravec (1998a) predicts that robots with computation power superior to that of the human brain will eventually supersede humans. The transhumanists, in contrast, do not anticipate that humanity will remain idle. Young's (2006) model of "harmonious complexification" (366) portrays life as moving toward increasing order, complexity, and self-organization and he sees humans as both producer and product of this process. Our species will initiate and ride the transhuman and posthuman wave. Kurzweil emphasizes the inevitability of this progression:

> [W]e are a product of evolution, indeed its cutting edge. But extending our intelligence by reverse engineering it, modeling it, simulating it, reinstantiating it on more capable substrates, and modifying and extending it is the next step in evolution. It was the fate of

bacteria to evolve into a technology-creating species. And it's our destiny now to evolve into the vast intelligence of the Singularity (298).

Transhumanists also describe evolution as if it were a steamroller. Evolution is the driving force of life and it crushes species that fail to keep pace with change. Given evolutionary pressure, the human species *must* move forward as a matter of survival. Furthermore, humans must continually adapt to sociotechnological change or else risk being left behind. To avoid enslavement or extinction, humans must become transhuman to stay ahead of robots or AI entities in computation power.

Homo Cyberneticus

A superior motivational system is not based solely on avoidance of harm; inspiration and commitment to a creative project are important as well. The transhumanists understand this and make sure to balance the concern for being displaced by machines with the promise of self-transformation. Evolution may be harsh and unforgiving but, according to the transhumanists, evolution has produced one species, homo sapiens, that is equipped and prepared to direct it. We are "steersmen," *Homo cyberneticus*, proclaims Simon Young. This is a bold declaration meant to inspire confidence and forward-thinking.

Clark (2003) explains that *we* are the creative project. In other words, because we are essentially dynamic and self-constructing and have the ability to expand with our technologies, we will continue to be the most innovative species.

> Our self-image as a species should not be that of ancient biological minds in colorful young technological clothes. Instead, ours are chameleon minds, factory-primed to merge with what they find and with what they themselves create (141).
>
> Our cognitive machinery is now intrinsically geared to self-transformation, artifact-based expansion, and a snowballing/bootstrapping process of computational and representational growth... Plasticity and multiplicity are our true constants (8).

The transition to transhumanity is not only manageable but is expected of "natural-born cyborgs."

Technological Momentum

In a paper published in 1965, Gordon Moore, co-founder of Intel, described how the number of transistors per square inch on integrated circuits had doubled every year between 1959 and 1965. He suggested that the trend might continue indefinitely. Moore's Law can be cautiously interpreted as a useful observation of past accomplishments or it can be understood as a predictor of future developments. Winner (1997) identifies a number of well-known writers including Nicolas Negroponte, George Gilder, Alvin Toffler, and Esther Dyson as expressing the latter view. Moravec (1998b) extends the trend line for MIPS (millions of

instructions per second) into the future to make the case that increase in computation power will allow for the rise of super-intelligent robots: "At the present rate,
computers suitable for humanlike robots will appear in the 2020s. Can the pace be
sustained for another three decades? The graph shows no sign of abatement." (6)

According to Kurzweil's (2001) "law of accelerating returns," the rate of
technological change is greater than commonly understood because

> technological change is exponential. In exponential growth, we find that a key measure-
> ment such as computational power is multiplied by a constant factor for each unit of time
> (e.g., doubling every year) rather than just being added to incrementally. Exponential
> growth is a feature of any evolutionary process, of which technology is a primary example.
> One can examine the data in different ways, on different time scales, and for a wide variety
> of technologies ranging from electronic to biological, and the acceleration of progress and
> growth applies. Indeed, we find not just simple exponential growth, but "double" expo-
> nential growth, meaning that the rate of exponential growth is itself growing exponen-
> tially. These observations do not rely merely on an assumption of the continuation of
> Moore's law (i.e., the exponential shrinking of transistor sizes on an integrated circuit), but
> is based on a rich model of diverse technological processes.

In his most expansive projections he takes for granted that exponential growth
will continue unabated and that "it will only take a quarter of a millennium (in our
own case) to go from sending messages on horseback to saturating the matter and
energy in our solar system with sublimely intelligent processes." (Kurzweil 2007)
Even the major social disturbances of the twentieth century have not delayed this
progression: "But the evolution of intelligence here on Earth is actually going very
well. All of the vagaries (and tragedies) of human history, such as two world wars,
the cold war, the great depression, and other notable events, did not make even the
slightest dent in the ongoing exponential progressions I previously mentioned."

It should be noted that Gordon Moore resists technological determinism, i.e.,
the view that technology is an autonomous force and proceeds according to
internal dynamics. He believes that social forces are pivotal to microchip advances, specifically, a collective-fulfilling prophecy is at work in which projected
trends become benchmarks for industry competition. Maybe it is a byproduct of
his macro level of analysis, but Kurzweil's law of accelerating returns does comes
across as being deterministic. He insists that exponential growth will be sustained
until it finally comes up against fundamental constraints of physics. Social constraints or, for that matter, social factors behind the construction and diffusion of
technologies receive little attention.

Being more cautious, other transhumanists make use of the ceteris paribus
clause, for instance, "if funding for research continues," "as long as we dispel
irrational fears," etc. Even so, they cite the same exponential trends and make the
same extrapolations as Kurzweil. Hughes (2004), for example, employs Moore's
Law as a template for growth rates in many fields and he predicts that

> NBIC technologies [nanotechnology, biotechnology, information technology and cogni-
> tive science] will definitely also change how we work, how we travel, how we commu-
> nicate, how we worship and how we cook. But the most fundamental changes in our lived
> experience will come from their impacts on our bodies and brains (7–8).

For Hughes it is not a question of *whether* technological advances will continue (they will), but how advances can be managed to best realize social values such as liberty and equality.

Technological acceleration, evolution, and *Homo cyberneticus* resound with the march of progress. This is a powerful theme in Western civilization that is now common to global culture. It is very difficult to deny or discredit. Critics may find it easier to dismiss the transhumanists' assertion that it will turn out good in the end than to dispel the common belief that there is no stopping change.

Conservationist Critique of Strong Claims

Religious Conservationist Counterargument

The religious critique of inevitability is based on the premise of the primacy of God's will. No alleged natural, human, or technological force is greater than that will, therefore none are truly immutable. It is possible that humans, with free will bestowed by the Creator, may choose a transhuman future, but they assert that this would be in defiance of God's plan. The creation account in Genesis is used to assert that humankind has been made in God's image and for the purpose of forming a special relationship with God, and that to radically alter the body or mind undermines this. (Herzfield 2002) Holding up *Homo cyberneticus* as *the* self-transforming, creative being constitutes the sin of pride and proclaiming the unstoppable progression of technologies is tantamount to idolatry. Catholic writers add transhuman alterations to a list of natural law violations that include abortion, contraception, and euthanasia. (Toth-Fejel 2004).

Hook poses a series of questions:

> Is it appropriate for members of the Body of Christ to engage in alterations that go beyond therapy and are irreversible? Is it just to do so in a world already deeply marked by inequities? What does it mean that our Lord healed and restored in his ministry—never enhanced? Is it significant that the gifts of the Holy Spirit—wisdom, love, patience, kindness—cannot be manufactured by technology? How would the transformation from *homo sapiens* to *techno sapiens* affect our identity as bearers of the image of God? (2004)

He believes that individuals (as well as communities and nations) retain free choice in the matter and should wrestle with these questions and not delay in making the right decision. Once transtechnologies are allowed to take off it will be that much more difficult to change course.

Finally, some fundamentalist Christians *accept* the inevitability of transhumanity as within God's plan. They expect that technology, society, and culture will become more inimical to true believers in the end times. Nevertheless, divine intervention and judgment will abolish transhumanity and God's Kingdom will be established on earth.

Secular Conservationist Counterargument

Secular conservationists charge that the transhumanists are guilty of reification to the extent they treat evolution, *Homo cyberneticus*, or technology as powerful autonomous agents when in reality they are constructs or, at the very most, processes that are co-determined by social factors. Agar (2004) takes Gould's approach and argues against trying to see in evolution a certain destiny. Gregory Peterson refutes "technological determinism"—a sense that technological change is beyond human control and is inevitable:

> It may be argued that there is little that we can do in any case, that the inevitable march of technology will triumph whether we oppose it or not. Once the automobile was invented, highways inevitably followed and those individuals who might have opposed them were powerless to stop the development. Such views of technological inevitability are misguided, for they miss the role that cultural context and communal will plays (2005).

I, too, am skeptical of the strong claims. I question whether the science of natural selection predicts socio-technological developments. I also question whether human nature can be reduced to an essence irrespective of particular society and culture. Finally, I understand technology to be part of sociotechnical ensembles and not an independent force. I find the moderate claims to be more persuasive, and so do many conservationists. They concede that social conditions and the cultural context favor transhumanity.

Moderate Claim: Social Conditions are Ripe

It would seem that if one truly believes that evolution, human nature, and/or technological momentum must bring about transhumanity there is no need to also make a moderate claim about conducive social conditions. Nevertheless, the moderate claim adds another dimension to the argument.

It is interesting that both proponents and opponents identify the same set of social conditions that favor transhumanity. Top on the list is nation-state competition. Cutting-edge science and engineering programs raise the cachet of countries and, more importantly, nations build such programs to promote economic and military interests. Post 9/11 it is especially difficult for conservatives in America to second guess the armed forces' strategy of maintaining technological superiority. Hook (2004), a conservative critic of transhumanism, admits as much when describing the possibility of transhuman soldiers: "The military feels a moral imperative to do whatever is necessary to make sure that each soldier comes home alive and well. If it takes genetic, cybernetic, or nanotechnological modifications to do that, so be it. After all, how could we deny our soldiers the greatest chance of survival?"

Conservationists on the right and left of the political spectrum show no such deference when it comes to the second enabler—capitalism. They complain that under this global economic system profit maximization drives production and consumption while ethical concerns are given short shrift. Mark Hanson asserts that

Enhancement technologies, too, are fueled by venture capitalists and companies who seek to capitalize on and even encourage dissatisfaction with the human conditions. And the consequence of these ventures are human traits offered as commodities—such as the products of many cosmetic surgeries and psychopharmacological agents—and distributed as the market demands...The confluence of these historical and cultural influences—along with certain other phenomena of popular culture that promote dissatisfaction with the finite body (such as media-driven conceptions of attractiveness and the so-called cult of celebrity)—generates fertile ground for the growth of human enhancement ambitions (1999).

The conservationists contend that governments have shown no appetite to limit them; indeed, with favorable patent laws, tax breaks, and industry-friendly regulatory systems, the opposite seems to be true.

Transhumanists have no problem with what they see as capitalism's penchant to add value to products and provide services to meet consumer demand. They expect global capitalism to promote transtechnologies. Showing a pragmatic side, Kurzweil (2001) asserts that

We will continue to build more powerful computational mechanisms because it creates enormous value. We will reverse-engineer the human brain not simply because it is our destiny, but because there is valuable information to be found there that will provide insights in building more intelligent (and more valuable) machines. We would have to repeal capitalism and every visage of economic competition to stop this progression.

Critics also contend that modern or postmodern culture lays the groundwork for permissiveness. In politics and law, liberalism provides a safe haven for self-development through the protection of personal freedoms. Bioethics is geared to the protection of individual rights. Whereas transhumanists have only praise for this, Kass (2002, p. 3) sees a dilemma:

The greatest dangers we confront in connection with the biological revolution arise not from principles alien to our way of life, but rather from those that are central to our self-definition and well-being: devotion to life and its preservation; freedom to inquire, invent or invest in whatever we want; a commitment to compassionate humanitarianism...

The culture and economy generate social competition which Sandel (2007) blames for driving the demand for enhancement technologies. We are most familiar with how this plays out in sports. Nearly all professional athletes need some assistance to compete at very demanding levels and they rely on trainers, therapists, dieticians, etc. Some make use of performance enhancing drugs (for example, amphetamines and anabolic steroids) in order to stay competitive or gain an edge. The World Anti-Doping Agency (WADA) has recently expressed concern over "gene doping" or the use of genetic technologies to boost performance. Sandel sees a similar trend with "hyperparenting." Today this entails private-school education, tutors, college admission coaches, and, if necessary, prescription drugs designed to focus attention or alleviate shyness. "Eugenic parenting" promises, at least theoretically, greater success because baseline capabilities may be augmented. He notes, however, that any competitive edge will be temporary as performance standards adjust upward, thereby prompting more dramatic enhancements.

The transhumanists, as well, anticipate this development, however, they accept social competition as the way of the world. Expanding on a celebrated argument from Adam Smith, they assert that overall wealth will increase as social competition drives personal innovation. For example, health, mental acuity, and personal productivity will improve as individuals "enhance to advance." Rather than see social competition as an exhausting, never-ending loop, they visualize a progressive spiral that moves individuals and societies forward.

The last favorable condition that I will mention has to do with the blurring of the social definition or distinction between therapy and enhancement. Therapy is the term applied to modifications that correct a pathology or deficit and enhancement is a term used to signify modifications that exceed a normal state. For example, a diabetic injects insulin for therapy whereas a perfectly healthy athlete injects anabolic steroids for enhancement. A clear distinction allows conservationists to single out enhancements for condemnation while supporting therapies and, in so doing, deflect an accusation of being indifferent to human suffering. Even better, they can accuse transhumanists of self-indulgence and deepening social injustice for prioritizing enhancement (which the affluent can afford) over therapy (which the poor need).

Holding the line is not as easy as it first appears and conservationists fear that it is becoming more untenable over time. For instance, therapies for some become enhancements for others. Human growth hormone, for example, is injected by those with and without the hormone deficiency. Research and development of therapies often provides a windfall for work on enhancements, thereby creating a practical dilemma for conservationists' strategy of containment. For instance, research on machine-brain interface primarily dedicated to addressing physical disability provides the basis for experimentation of new abilities.

Maintaining a clear distinction between therapy and enhancement is difficult in that social conventions for what is normal and pathological are like sand dunes that shift over time. Consider these examples: orthodontics problematized crooked teeth and cosmetic surgery, botox, and facial lotions have done the same for wrinkles. Vision and hearing loss associated with aging are deemed impairments and are corrected through surgery and aids. Pharmaceutical corporations advertise drug therapies to adjust temperaments or moods that in the past were considered within a normal range. New treatments for decline in memory, strength, agility, etc., could prompt a change in attitude toward senescence so that it is perceived as a disease. If so, today's longevity enhancements will be tomorrow's therapies. As Hanson (1999, p. 125) notes,

> Just as medical technologies now increasingly redefine sickness in terms of what medical technologies are able to diagnose and treat… what is now considered healthy or normal will increasingly be thought of as defective or disvalued as enhanced states become more the norm. In other words, what was once "normal" or "healthy" becomes something from which we suffer. The irony of enhancement technologies is that its very success serves to broaden the scope of conditions from which humans can be said to suffer.

Baylis and Robert (2004, p. 17) point out that affluent nations have a poor track record halting enhancement technologies. Despite prohibitions, performance enhancing drugs and illicit mind-altering drugs are widely available. Neuropharmacology, organ transplantation, gender reassignment, cosmetic surgery, and human growth hormone treatment are sanctioned under medicine. Public reaction to enhancement technologies becomes muted over time from "initial condemnation, followed by ambivalence, questioning and limited use, followed in turn by a change in public perceptions, advocacy and widespread acceptance."

Relinquishment

To mount a successful opposition campaign, conservationists must persuade the general public and the powers that be to give up promising technologies. However, it is unusual for those with the means to afford medical and technological advancements to forgo them on principle. Relinquishment is an option that few choose.

Joy (2000) takes a different approach with his case for relinquishment by drawing attention to the unprecedented dangers of GNR technologies (genetics, nanotechnology, and robotics), most importantly the escape or release of self-replicating entities. In the "grey goo" scenario, for example, nanobots that derive their energy from the sun multiply so fast that they coat vegetation and interrupt photosynthesis. Joy advocates relinquishment to prevent cataclysm. He provides examples of governments exercising such will and relinquishing dangerous technologies as with the signing of the 1972 Biological Weapons Convention and the 1993 Chemical Weapons Convention. This demonstrates that international cooperation is possible and that bans can work.

However, if the threat of runaway replicators is reduced or removed through research and development protocols, state regulation, etc., would the public demand relinquishment of GNR technologies? In other words, if the public could be persuaded that catastrophe is unlikely, would they support banning technologies that could provide personal and social benefits? So many medical and cosmetic products, such as prescription drugs and cosmetic surgery, have both positive *and* negative effects and although not everyone elects to use these products there is no groundswell to ban them.

In his seminal work, *Risk Society* (1992), Beck points out that modernization represents in the minds of its recipients a tradeoff between comfort and risks. For example, to have air conditioning, suburban enclaves, and economic growth we consume more energy and in the process run the risk of global warming. Mass transportation and international travel increase the risk of terrorist attacks and viral epidemics. Mass production of food entails the passage of pesticides, growth hormones, and antibiotics into our diet. Civilization has come to be associated with a level of endangerment. Many people accept that progress has a price, and it is worth paying. If enhancement technologies are understood in terms of progress, and not in terms of weapons of mass destruction, the case for relinquishment will be a hard sell.

Chapter 6
Closure

When will the transhumanity debate end? How will closure be accomplished? We know that social debates end when real world conditions change or interest wanes. For example, slavery was highly controversial in America during the eighteenth and nineteenth centuries but is no longer today. Scientific controversies over competing theories also get settled as, for instance, when Einstein's theory of general relativity superseded Newton's theory of gravity. Power is always exercised in closures and this was most evident in the slavery debate when the central government used force and authority during the Civil War and Reconstruction. Although insufficient to settle the slavery debate, persuasion is a form of power and has been pivotal in many scientific revolutions. It is important, however, not to overstate differences between how social and scientific debates conclude. Scientists can always find a way to contest theory and evidence. Closure sometimes requires a major shakeup in the scientific establishment.

No Easy Resolution

Given that the transhumanity debate is a contest that entails a nest of thorny social, political, and ethical issues, I very much doubt that persuasion will suffice to end this controversy. Most importantly, the contestants do not agree on principles, desirable ends, acceptable risks, methodology, and proof of evidence. Should the Bible be treated as an authoritative text? Is social change preferable to tradition? Can we trust facts purported by corporate science? Without consensus on these and other critical points there can be no mutually agreed set of ground rules by which to judge the arguments.

Also, I don't see the transhumanity debate dissipating as do fads. Barring a major catastrophe that would seriously impair global capitalism, I expect increasing technological innovation and the encouragement of its consumption. The line between therapy and enhancement will be tested in more ways. I see no abatement of the culture wars and I expect the population will remain divided over enhancement

S. Lilley, *Transhumanism and Society*, SpringerBriefs in Philosophy, DOI: 10.1007/978-94-007-4981-8_6, © The Author(s) 2013

technologies. Social movement leaders, politicians, and news reporters will continue to see advantages in stoking the debate. If my forecast is accurate, the number of individuals drawn into the debate will grow rather than diminish.

Transhumanists predict success, but I am skeptical of the more deterministic accounts. For Ray Kurzweil, the debate soon will be moot as transhumanity takes off in accordance with the law of accelerating returns and the evolution of intelligence. However, even if scientific discovery and technological innovation continue at a rapid pace, applications for the body will have to be tested and approved under a review and regulatory system that is comparatively glacially slow. It is this rate-limiting factor that will dictate the pace of change. Also, funding for research and development of controversial technologies may be delayed or blocked in times when the allies of conservationists have more influence in government. The general point that I am making is that political, economic, and social factors will get in the way. We should expect stops and starts, bottlenecks, and perhaps reversals and chokepoints for enhancement technologies, and every skirmish and battle will reinvigorate the debate.

Balancing Act with Inevitability Claims

As discussed in the previous chapter, claims of inevitability are made for social and political reasons. Even if individuals are uneasy about the prospects of a transhuman future, they will not rise in opposition if resigned to it. It is in transhumanists' best interests to press the inevitability claim, however, if done too emphatically this creates two problems. First, prospective *supporters* may conclude that they need not become actively engaged in the movement. Why get involved and make sacrifices when the outcome is inevitable? Passivity may be a desirable disposition for potential opponents, but not for supporters! At the very least such support might prove crucial in the near term to meet potential political and regulatory challenges. Second, the more detailed and imminent the transhumanity predictions, the more captivating the inevitability claim. However, specific predictions become contested points. Supporter confidence is shaken and opponents emboldened when expected milestones are not met. Here again, Ray Kurzweil's timetable regarding the rapid inception and diffusion of transtechnologies is problematic.

If we consider the back-and-forth policy on embryonic stem cell research in the United States and if we take into account that clinical trials average 6 years to complete, there is reason to believe that therapeutic or human enhancement technologies will not match the rate of rapid development of other technologies, for example, in computer chip manufacturing. Conceding that social forces will influence the timetable reduces the risk of making incorrect predictions and provides prospective supporters reason for activism. This more pragmatic approach is recommended by James Hughes.

Conservationists have their own balancing act to perform. To provoke a sense of urgency in their respective base and in the general public they must present a compelling case that a point of no return soon will be reached. They identify an alignment of powerful social forces that drive transtechnology innovation forward and assert that, absent a bold stand for relinquishment, enhancement technologies will take off, lifted by social competition, change in norms, etc. However, this may have the opposite intended effect and foster fatalism. Some may wonder whether an anti-transhumanity campaign is quixotic. Being on high alert is exhausting. Can vigilance be sustained? Social movements often fail as supporters run out of energy and are drawn back to ordinary demands of work and family. On the other hand, lowering the alarm increases complacency. Finding the optimum level is difficult.

Conservationists must also wrestle with how to pitch optimism. A certain amount of hope is necessary but overconfidence leads to complacency. Conservationists strike a balance by holding up *hard won* victories. For instance, they point to the international bans on biological and chemical weapons to provide supporters with hope that transtechnologies may also be stopped, all the while reminding supporters that comprehensive bans are won through proponents' hard work and dedication.

Scenarios

Both sides in the transhumanity debate point to success stories, but like the risk objects previously discussed, the victories are about sociotechnological contests once removed, e.g., nuclear test ban treaties, fluoridation, and immunization. As with risk rhetoric, the "rhetoric of success" at this point is relatively unencumbered by facts on the ground. Of course, this will change as the years go by and the pace of development is made evident. Proposals for clinical trials will be accepted or rejected, providing clear wins or losses. Transhumanists will emphasize promising results and conservationists will use negative results to bolster their case for relinquishment.

At some point a string of victories and favorable facts will make the transhumanists' case so compelling that a majority will accept it, opening a clear path for transtechnologies. In a future age of transhumanity it will be taken for granted that success was inevitable. Or, at some point a string of defeats or disquieting facts will bolster the conservationists' case and they will garner sufficient resistance to ban transtechnologies.

Of course, it is quite possible that there will be no clear resolution for the foreseeable future. Conservationists and transhumanists insist that we are at a crossing point and they would have us believe that soon there will be an unequivocal winner or loser in this contest. Hyperbole is useful in attracting attention and establishing the frame of the debate, however the stark all-or-nothing imagery may be misleading. If instead of total victory or defeat, what if the future for enhancement technologies is muddled?

Radical transformation is a wonder to contemplate and, depending on one's perspective, very intriguing or disquieting. Nevertheless, *moderate* transformation must precede it because the science and engineering of radical transformation can only mature through initial exploratory studies utilizing human subjects.

I expect the debate to enter a second stage when applications that bend or blur the therapy/enhancement line are tested on humans. Conservationists and their allies will mount a vigorous and protracted risk contest. Even so, resources are finite. Facts will emerge from experiments regarding benefits and deleterious side effects, and conservationists may decide for tactical reasons to marshal resources to undermine the most vulnerable applications. Transhumanists and their allies may decide to cut their losses on such and focus their resources on promoting the applications with the best benefit-risk profile. Will all applications come to fruition, will all be relinquished? In all probability, such extremes are less likely to occur than a mixed outcome of successes and failures.

We can't even assume that the present alignment of contestants will remain so neat. New players may enter the debate and some contestants may switch sides. It has been the case so far that when religious conservatives and transhumanists argue, a secular-faith divide is pronounced. However, we should not assume from this that others drawn into the debate could not or would not accommodate their religious values to enhancement technologies. Innovators will see the advantage in being the first to open a space for integration. If successful, others will follow suit and the debate will no longer cleanly fall on the secularization fault line.

Factions develop in social movements and new leaders often propose new agendas. For example, a few prominent figures in the environmental movement have expressed support for the inclusion of nuclear energy in a plan to reduce global warming and some feminists have questioned the women's movement's condemnation of pornography. Transhumanists seem perplexed that liberation movements have not been more sympathetic to transhumanism, but we shouldn't rule out the possibility that in the future a vanguard will embrace some enhancement technologies.

Assuming that at least some enhancement technologies succeed we are likely to see a divergent utilization pattern in legal and black market use. I agree with Gregory Stock that consumers will opt for safe, reliable, and reversible procedures over those that are exotic, risky, and permanent. This suggests incremental change and certainly not a great leap from human to transhuman. I expect the population use pattern to approach a normal distribution with a small percentage rejecting all applications, a small percentage incorporating many applications and the majority selecting a modest set. The rhetoric contest in this third stage of the transhumanity debate will be about how augmentations play out in the lives of individuals and in the social world: Physical and emotional improvement or impairment? Consensus, competition, or social conflict? Increase or decrease in social problems? C. Wright Mills' [(1959) 1969] questions will be asked of the enhanced: "In what ways are they selected and formed, liberated and repressed, made sensitive and blunted?"

The transhumanity debate at this point in time is a contest over values, ideas, and imagined futures. Intellectuals and specialists are the main contestants. Their arguments will be relevant in later stages of the transhumanity debate, however, everything else about the contest will change. I don't see any easy way out of the transhumanity debate. It will close only after a time when populations are exposed to *trans*sociotechnical ensembles. The general public will be drawn in—not for the chance to argue philosophy, social theory, science and religion but in order to decide whether to take the plunge and open their minds and bodies and that of their children to enhancement technologies. They will do so in the face of advertising/ marketing campaigns, novel social movements, social competition, peer pressure, and strident politics.

Now the transhumanity debate is about the ideal path, but by the third stage it increasingly will be about the lived response. I can't imagine all the many demands and distractions that we may face, although I believe, to paraphrase Karl Marx, we will accept or reject transhumanity but not in circumstances of our own choosing. This has been true of all momentous revolutions and I wouldn't expect otherwise from this one.

About the Author

Stephen Lilley (PhD and MA, University of Massachusetts, Amherst; BA, College of the Holy Cross) teaches and conducts research on society and technology. He has written on genetic medicine, Moore's Law, characterizations of human-machine networks, and technology fatalism. He has co-authored studies and papers on social networking, surveillance, and privacy. His writing on social movements began with a Master's thesis on liberation theology in the Sandinista Revolution and a dissertation on Dr. King's leadership within the Civil Rights Movement in America. More recently he has studied the ideology, rhetoric, and mobilization strategies of organizations that strive to advance controversial technologies. He is a Professor of Sociology at Sacred Heart University, Fairfield CT.

S. Lilley, *Transhumanism and Society*, SpringerBriefs in Philosophy, DOI: 10.1007/978-94-007-4981-8, © The Author(s) 2013

References

Agar, N. (2004). *Liberal eugenics: In defense of human enhancement*. Malden, MA: Blackwell Publishing.

Annas, G. (2006). Governing biotechnology. *Global Agenda,1*(4), 21.

Bailey, R. (2005). *Liberation biology: The scientific and moral case for the biotech revolution*. Amherst, MA: Prometheus Books.

Bainbridge, W. S. (2006). Cyberimmortality: Science, religion, and the battle to save our souls. *Futurist,40*(2), 25–29.

Baylis, F., & Robert, J. S. (2004). The inevitability of genetic enhancement technologies. *Bioethics,18*(1), 1–26.

Beck, U. (1992). *Risk society: Towards a new modernity* (M. Ritter, Trans.). London: Sage Publications.

Beck, U. (1995). *Ecological politics in an age of risk* (A. Weisz, Trans.). Cambridge, UK: Polity Press.

Becker, B. (2000). Cyborgs, agents, and transhumanists. *Leonardo,33*(5), 361–366.

Bedsworth, L. W., Lowenthal, M. D., & Kastenberg, W. E. (2004). Uncertainty and regulation: The rhetoric of risk in the California low-level radioactive waste debate. *Science, Technology, & Human Values,29*(3), 406–427.

Bendle, M. F. (2002). Teleportation, cyborgs and the posthuman ideology. *Social Semiotics, 12*(1), 45–62.

Benedict, XVI. (2007). *Spe Salvi*. Retrieved on March 4, 2008 (http://www.vatican.va/holy_father/benedict_xvi/encyclicals/documents/hf_ben-xvi_enc_20071130_spe-salvi_en.html).

Bergsma, Ad. (2000). Transhumanism and the wisdom of old genes: Is neurotechnology a source of future happiness? *Journal of Happiness Studies,1*(3), 401–417.

Berry, R. J. (2006). Fabricated nature: Where are the boundaries? *Ecotheology: Journal of Religion, Nature & the Environment,11*(1), 9–31.

Bijker, W. E. (1995). *Of bicycles, bakelites, and bulbs: Toward a theory of sociotechnical change*. Cambridge, MA: MIT Press.

Bijker, W. E., & Pinch, T. J. (1984). The social construction of facts and artefacts: Or how the sociology of science and the sociology of technology might benefit each other. *Social Studies of Science,14*(3), 399–441.

Blakeslee, S. (2008). Monkey's thoughts propel robot, a step that may help humans. *New York Times*, January 15. Retrieved January 15, 2008 (http://www.nytimes.com/2008/01/15/science/15robo.html?ref=science).

Block, J. L. (2008). Issues for DSM-V: Internet addiction. *American Journal of Psychiatry,165*(3), 306–307.

Bostrom, Nick. (2005). In defense of posthuman dignity. *Bioethics,19*(3), 202–214.

Cage, S. (2008). High prices nudge Europe nearer to GM food. *Reuters* July 7, 2008. Retrieved July 7, 2008 (http://www.reuters.com/article/inDepthNews/idUSL0256084320080708?feed Type=RSS&feedName=inDepthNews&rpc=22&sp=true).

Campbell, N. (2005). Suspect technologies: Scrutinizing the intersection of science, technology, and policy. *Science, Technology, & Human Values,30*(3), 374–402.

Chislenko, A. (1995). Legacy systems and functional cyborgization of humans. Retrieved March 5, 2009 (http://www.lucifer.com/~sasha/articles/Cyborgs.html).

Clark, A. (2003). *Natural-born cyborgs: Minds, technologies, and the future of human intelligence.* New York: Oxford University Press.

Clive, J. (2009). *Global Status of Commercialized Biotech/GM Crops: 2009.ISAAA BriefNo. 41.* ISAAA: Ithaca, NY.

Clynes, M. E., & Nathan, S. K. (1960). Cyborgs and space. *Astronautics*, September issue: 26–27 and 74–75.

Collins, H., & Pinch, T. (1993). *The golem: What everyone should know about science.* Cambridge: Cambridge University Press.

Corea, Gina. (1986). *The mother machine: From artificial insemination to artificial wombs.* New York: Harper and Row.

Davis, E. (1998). *Techgnosis: Myth, magic, and mysticism in the age of information.* London: Serpent's Tail.

Davis, J. J. (2004). Christian reflections on the genetic revolution. *Evangelical Review of Theology,28*(1), 65–79.

DeLashmutt, M. W. (2006). Perspectives on techno-science and human nature. *Zygon: Journal of Religion & Science,41*(2), 267–287.

de Magalhaes, J. P. (2002). The one-man rule. *Futurist,36*(6), 41–45.

Edge. (2006). The edge annual question—2006: What is your dangerous idea? Retrieved June 4, 2009 (http://www.edge.org/q2006/q06_index.html).

Elliott, C. (2003). Humanity 2.0. *Wilson Quarterly, 27*(4), 13.

FINRRAGE. (2008). Retrieved June 16, 2008 (http://www.finrrage.org/index.html).

Frodeman, R. (2006). Nanotechnology: The visible and the invisible. *Science as Culture,15*(4), 383–389.

Fukuyama, F. (2002). *Our posthuman future: Consequences of the biotechnology revolution.* New York: Farrar, Straus and Giroux.

Geertsema, H. G. (2006). Cyborg: Myth or reality? *Zygon: Journal of Religion & Science,41*(2), 289–328.

Gould, S. J. (2007). In S. Rose (Ed.), *The richness of life: The essential Stephen Jay Gould.* New York: WW Norton.

Graham, E. (2006). In whose image? Representations of technology and the 'ends' of humanity. *Ecotheology,11*(2), 159–182.

Great GO. (2006). The political becomes personal. *Earth First! Journal.* Retrieved January 3, 2008 (http://www.earthfirstjournal.org/index.php).

Habermas, J. (2003). *The future of human nature* (W. Rehg, M. Pensky & H. Beister, Trans.). Malden, MA: Blackwell Publishing.

Hanson, M. J. (1999). Indulging anxiety: Human enhancement from a protestant perspective. *Christian Bioethics,5*(2), 121–138.

Hayles, N. K. (1999). *How we became posthuman: Virtual bodies in cybernetics, literature, and informatics.* Chicago, Illinois: University of Chicago Press.

Heidegger, M. [1954] (2003). The question concerning technology. In R. C. Scharff & V. D. Malden (Eds.), *Philosophy of technology: The technological condition: An anthology* (pp. 252–264). MA: Blackwell Publishing.

Herzfield, N. (2002). *In our image: Artificial intelligence and the human spirit.* Minneapolis, MN: Fortress Press.

Hook, C. C. (2004). The techno sapiens are coming. *Christianity Today Magazine,48*(1), 36–41.

Hughes, J. (2004). *Citizen cyborg: Why democratic societies must respond to the redesigned human of the future*. Cambridge, MA: Westview Press.

Humanity +. (2010). Transhumanity declaration. Retrieved July 13, 2010 (http://humanityplus.org/learn/transhumanist-declaration/).

Huxley, A. [1932] (1969). *Brave new world*. New York: Harper & Row.

Huxley, J. (1957). *New bottles for new wine*. London: Chatto & Windus.

International Narcotics Control Board. (2006). *Report of the international narcotics control board for 2006*. Retrieved September 5, 2007 (http://www.incb.org/incb/en/annual_report_2006.html).

Jasanoff, S. (1995). Product, process, or programme: Three cultures and the regulation of biotechnology. In M. Bauer (Ed.), *Resistance to new technology* (pp. 311–331). Cambridge: Cambridge University Press.

John Paul II. (1981). *Laborem* Exercens. Retrieved April 10, 2007 (http://www.vatican.va/holy_father/john_paul_ii/encyclicals/documents/hf_jp-ii_enc_14091981_laborem-exercens_en.html).

Joy, B. (2000). Why the future doesn't need us. *Wired Magazine,8*(4), 238–262.

Junker-Kenny, M. (2005). Genetic enhancement as care or as domination? The ethics of asymmetrical relationships in the upbringing of children. *Journal of Philosophy of Education,39*(1), 1–17.

Kass, L. (2002). *Life, liberty and the defense of dignity: The challenge for bioethics*. San Francisco: Encounter Books.

Kim, B. N. (2007). From internet to "Family-Net": Internet addict vs. digital leader. In *International Symposium on the Counseling and Treatment of Youth Internet Addiction* (p. 196). Seoul, Korea: National Youth Commission.

Kuhn, T. (1962). *The structure of scientific revolutions*. Chicago: University of Chicago Press.

Kurzweil, R. (2001). The law of accelerating returns. Retrieved July 29, 2008 (http://www.kurzweilai.net/meme/frame.html?main=/articles/art0134.html).

Kurzweil, R. (2005). *The singularity is near: When humans transcend biology*. New York: Penguin.

Kurzweil, R. (2007). Foreward to the intelligent universe. Retrieved July 29, 2010 (http://www.kurzweilai.net/forward-to-the-intelligent-universe).

Kurzweil, R., & Goodman, T. (2009). *Transcend: Nine steps to living well forever*. New York: Rodale Press.

Lassen, J., & Jamison, A. (2006). Genetic technologies meet the public: The discourses of concern. *Science, Technology, & Human Values,31*(1), 8–28.

Latour, B. (1992). Where are the missing masses? The sociology of a few mundane artifacts. In W. E. Bijker & J. Law (Eds.), *Shaping technology/building society* (pp. 225–258). Cambridge: MIT Press.

Lauritzen, P. (2005). Stem cells, biotechnology, and human rights: Implications for a posthuman future. *Hastings Center Report,35*(2), 25–33.

Lee, S.-J., & Chae, Y.-G. (2007). Children's internet use in a family context: Influence on family relationships and parental mediation. *CyberPsychology & Behavior,10*(5), 640–644.

Lilley, S. (2007). Catholic students' fatalism in anticipation of transhuman technologies. *International Journal of Interdisciplinary Social Sciences,2*(1), 313–319.

Lyons, D. (2009). I, robot: One man's quest to become a computer. *Newsweek,153*(21), 66–73.

McKibben, B. (2003). *Enough: Staying human in an engineered age*. New York: Henry Holt and Company.

McKibben, B. (2004). Helping hand or big fat fist? Why countries are saying no thanks to U.S. food aid. Retrieved July 14, 2008 (http://www.mindfully.org/GE/2004/Zambia-US-Food-Aid1jul04.htm).

Meenan, A. L. (2007). Internet gaming: A hidden addiction. *American Family Physician,76*(8), 1117.

Mellman Group. (2006). *Public sentiment about genetically modified food (2006)*. Pew initiative on food and biotechnology. Retrieved May 27, 2008 (http://www.pewtrusts.org/news_room_detail.aspx?id=32802).

Mills, C. W. [1959] (1969). *The sociological imagination*. New York: Oxford University Press.

Moore, G. F. (1965). Cramming more components onto integrated circuits. *Electronics, 38*(8), 114–117. Retrieved June 12, 2006 (ftp://download.intel.com/museum/Moores_Law/Articles-Press_Releases/Gordon_Moore_1965_Article.pdf).

Moravec, H. (1998a). *Mind children: The future of robot and human intelligence*. Cambridge, MA: Harvard University Press.

Moravec, H. (1998b). When will computers match the human brain? *Journal of Evolution and Human Technology* 1(1). Retrieved September 10, 2008 (http://www.transhumanist.com/volume1/moravec.htm).

More, M. [2000] (2006). Embrace, don't relinquish the future. In M. E. Winston & R. D. E. Belmont (Eds.), *Society, ethics, and technology*, 3rd ed. (pp. 239–244). CA: Wadsworth/Thomson.

More, M. (2003). Principles of extropy (version 3.11). Retrieved on January 28, 2008 (http://www.extropy.org/principles.htm).

More, M. (2005). The proactionary principle (version 1.2). Retrieved on January 29, 2008 (http://www.maxmore.com/proactionary.htm).

National Center for Health Statistics. (2007). *Health, United States, 2007, with chartbook on trends in the health of Americans*. Hyattsville, MD: US Government Printing Office.

Nie, N. H., & Hillygus, D. S. (2002). The impact of internet use on sociability: Time-diary findings. *IT & Society,1*(1), 1–20.

Noble, D. F. [1997] (1999). *The religion of technology: The divinity of man and the spirit of invention*. New York: Penguin Group.

O'Brien, M. (2001). *The politics of reproduction*. New York: HarperCollins Publisher.

Padgett, A. G. (2005). God versus technology? Science, secularity, and the theology of technology. *Zygon: Journal of Religion & Science,40*(3), 577–584.

Parthasarathy, S. (2004). Regulating risk: Defining genetic privacy in the United States and Britain. *Science, Technology, & Human Values,29*(3), 332–352.

Pearce, D. (1998). The hedonistic imperative. Retrieved April 4, 2006 (http://www.hedweb.com/).

Peterson, G. R. (2005). Imaging god: Cyborgs, brain-machine interfaces, and a more human future. *Dialog: A Journal of Theology,44*(4), 337–346.

Powell, C. S. (2000). Twenty ways the world could end suddenly. *Discover Magazine,21*(10), 50–57.

Rabino, I. (2003). Genetic testing and its implications: Human genetics researchers grapple with ethical issues. *Science, Technology, & Human Values,28*(3), 365–402.

Rifkin, J. (2005). Ultimate therapy: Commercial eugenics in the 21st century. *Harvard International Review,27*(1), 44–48.

Rowland, R. (1992). *Living laboratories: Women and reproductive technologies*. Bloomington: Indiana University Press.

Sandel, M. J. (2007). *The case against perfection: Ethics in the age of genetic engineering*. Cambridge, MA: Harvard University Press.

Schweiker, W. (2003). Theological ethics and the question of humanism. *Journal of Religion,83*(4), 539–561.

Selin, C. (2007). Expectations and the emergence of nanotechnology. *Science, Technology, & Human Values,32*(2), 196–220.

Sjöberg, L. (2002). The allegedly simple structure of experts' risk perception: An urban legend in risk research. *Science, Technology, & Human Values,27*(4), 443–459.

Song, R. (2006). Knowing there is no god, still we should not play god? Habermas on the future of human nature. *Ecotheology: Journal of Religion, Nature & the Environment,11*(2), 191–211.

Star, S. L. & Griesemer, J. R. [1989] (1999). Institutional ecology, "Translations" and boundary objects: Amateurs and professionals in Berkeley's museum of vertebrate zoology, 1907–1939. In M. Biagioli (Ed.), *The science studies reader* (pp. 503–524). New York: Routledge.

Stock, G. (2002). *Redesigning humans: Our inevitable genetic future*. Boston: Houghton Mifflin Company.

Stock, G. (2003). From regenerative medicine to human design: What are we really afraid of? *DNA and Cell Biology,22*(11), 679–683.

Sullins, J. (2000). Transcending the meat: Immersive technologies and computer mediated bodies. *Journal of Experimental & Theoretical Artificial Intelligence,12*(1), 13–22.

Szerszynski, B. (2006). Techno-demonology: Naming, understanding and redeeming the A/human agencies with which we share our world. *Ecotheology: Journal of Religion, Nature & the Environment,11*(1), 57–75.

Tapia, A. H. (2003). Technomillennialism: A subcultural response to the technological threat of Y2 K. *Science, Technology, & Human Values,28*(4), 483–512.

Tindale, C. W. (2004). *Rhetorical argumentation: Principles of theory and practice*. Thousand Oaks, CA: Sage Publications.

Toth-Fejel, T. T. (2004). Humanity and nanotechnology: Judging enhancements. *National Catholic Bioethics Quarterly,4*(2), 335–364.

Turkle, S. (1995). *Life on the screen: Identity in the age of the internet*. New York: Simon & Schuster.

Universal Declaration of Human Rights. (2008). G.A. res. 217A (III), U.N. Doc A/810 at 71.

Warwick, K. (2008). Frequently asked questions. Retrieved March 28, 2008 (http://www. kevinwarwick.com/faq.htm).

Wetmore, J. M. (2004). Redefining risks and redistributing responsibilities: Building networks to increase automobile safety. *Science, Technology, & Human Values,29*(3), 377–405.

Wingspread Conference on the Precautionary Principle. (1998). *The wingspread consensus statement on the precautionary principle*. Retrieved June 26, 2008 (http://www.sehn. org/wing.html).

Winner, L. (1997). Cyberlibertarian myths and the prospects for community. *Computers and Society,27*(3), 14–19.

Woliver, L. R. (2002). *The political geographies of pregnancy*. Urbana, IL: University of Illinois Press.

World Transhumanist Association. (2008). Retrieved April 24, 2008 (http://www.trans humanism.org/index.php/WTA/index/).

Wynne, B. (1987). *Risk management and hazardous waste: Implementation and the dialectics of credibility*. Berlin: Springer.

Young, S. (2006). *Designer evolution: A transhumanist manifesto*. Amherst, NY: Prometheus Books.

Zald, M. N., & McCarthy, J. D. (1979). *Social movements in an organizational society*. New Brunswick, NJ: Transaction Books.

Zaracostas, J. (2007). Misuse of prescription drugs could soon exceed that of illicit narcotics, UN panel warns. *British Medical Journal,334*(7591), 444.

Index

S. Lilley, *Transhumanism and Society*, SpringerBriefs in Philosophy,
DOI: 10.1007/978-94-007-4981-8, © The Author(s) 2013

Lightning Source UK Ltd.
Milton Keynes UK
UKOW06f1800050416

271622UK00003B/155/P